Multilabel Classification

Francisco Herrera · Francisco Charte
Antonio J. Rivera · María J. del Jesus

Multilabel Classification

Problem Analysis, Metrics and Techniques

 Springer

Francisco Herrera
University of Granada
Granada
Spain

Francisco Charte
University of Granada
Granada
Spain

Antonio J. Rivera
University of Jaén
Jaén
Spain

María J. del Jesus
University of Jaén
Jaén
Spain

ISBN 978-3-319-82269-3   ISBN 978-3-319-41111-8   (eBook)
DOI 10.1007/978-3-319-41111-8

Printed on acid-free paper

This Springer imprint is published by Springer Nature
The registered company is Springer International Publishing AG Switzerland

To my family

Francisco Herrera

To María Jesús, my beloved life partner

Francisco Charte

To my family

Antonio J. Rivera

To Jorge, my beloved life partner

María J. del Jesus

Preface

The huge growth of information stored everywhere, from mobile phones to datacenter servers, as well as the large user base of many Internet services, such as social networks and online services for publishing music, pictures, and videos, demands automated systems for categorizing and labeling all this information. A common characteristic of texts published in news sites and blogs, videos, images, and pieces of music is that all of them can be assigned to multiple categories at once. Hence, the need to have algorithms able to adequately classify the data assigning it the proper labels.

Multilabel classification is a data mining area that encompasses several tasks specific for this type of data, including custom metrics aimed to characterize multilabel datasets and also to evaluate results, specialized preprocessing methods able to solve the peculiarities of multilabeled data, and also specific classification algorithms qualified for learning from this type of data, among others. Most of these techniques are pretty new and many of them are still in development.

Multilabel classification is a topic which has generated a notable interest in late years. Beside its multiple applications to classify different types of online information, it is also useful in many other areas, such as genomic and biology. Consequently, the demand for multilabel techniques is constantly growing. This book will guide the reader to the discovery of all aspects of multilabel classification.

Based on the experience of the authors after several years focused on multilabel learning techniques, this book reviews the specificities of this kind of classification, including all the custom metrics and techniques designed to deal with it, and provides a comprehensive reference for anyone interested in the field.

After portraying the context that multilabel classification belongs to, in the introduction, a formal definition of this problem along with a broad view on how it has been faced and the fields it has been applied to are provided in the second chapter. The third one is devoted to introducing most of the publicly available multilabel use cases, as well as the metrics defined to characterize and evaluate them. Chapters 4–6 review multilabel classification methods grouping them into three groups, depending on the approach followed to tackle the task, data transformation, method adaptation, or the use of ensembles. Two of the most relevant obstacles in working with multilabel data, high dimensionality and class imbalance, are discussed in Chaps. 7 and 8. Chapter 9 introduces several software tools and frameworks aimed to ease the work with multilabel data, including obtaining this kind of datasets, performing exploratory analysis and conducting experiments.

Although multilabel learning is still in an early development stage with respect to other data mining techniques, the amount of proposed algorithms, most of them classification methods, is impressive. In the foreseeable future, it predictably will further expand to additional application fields, and the volume of new techniques grows almost every day.

The intended audience of this book are developers and engineers aiming to apply multilabel techniques to solve different kinds of real-world problems, as well as researchers and students needing a comprehensive review on multilabel literature, methods, and tools. In addition to the text itself, the authors supply the readers with a software repository containing data, code, and links, along with two R packages as tools to work with multilabel data.

We wish to thank all our collaborators of the research groups "Soft Computing and Intelligent Information Systems" and "Intelligent Systems and Data Mining." We are also thankful to our families for their helpful support.

Granada, Spain   Francisco Herrera
Granada, Spain   Francisco Charte
Jaén, Spain   Antonio J. Rivera
Jaén, Spain   María J. del Jesus
May 2016

Contents

Acronyms

ACO   Ant colony optimization
ADT   Alternative decision trees
ANN   Artificial neural network
API   Application programing interface
ARFF   Attribute-Relation File Format
AUC   Area under the ROC curve
BCC   Bayesian classifier chains
BID   Binary datasets
BoW   Bag of words
BR   Binary relevance
CC   Classifier chains
CCA   Canonical correlation analysis
CDE   ChiDep ensemble
CL   Compressed labeling
CLR   Calibrated label ranking
CML   Collectible multilabel
CMLPC   Calibrated pairwise multilabel perceptron
CRAN   Comprehensive R Archive Network
CRF   Conditional random fields
CS   Compressed sensing
CSV   Comma-separated values
CT   Classifier trellis
CV   Cross-validation
CVIR   Coefficient of variation for the average imbalance ratio (*MeanIR*)
CVM   Core vector machine
DLVM   Dual-layer Voting Method
DM   Data mining
DT   Decision trees
ECC   Ensemble of classifier chains
ELM   Extreme learning machine
EML   Ensemble of multilabel learners

EPS	Ensemble of pruned sets
FN	False negatives
FP	False positives
IBL	Instance-based learning
IR	Imbalance ratio or information retrieval depending on the context
JDK	Java Development Kit
JRE	Java Runtime Environment
KDD	Knowledge discovery in databases
KDE	Kernel dependency estimation
kNN	k-nearest neighbors
LDA	Linear discriminant analysis
LP	Label powerset
LSI	Latent semantic indexing
MAP	Maximum a posteriori probabilities
MCD	Multiclass datasets
MIR	Mean imbalance ratio
MLC	Multilabel classification
MLD	Multilabel dataset
MLP	Multilayer perceptron
OVA	One-vs-all
OVO	One-vs-one
PCA	Principal component analysis
PCC	Probabilistic classifier chains
PCT	Predictive clustering tree
PMM	Probabilistic mixture models
PS	Pruned sets
PSO	Particle swarm optimization
QCLR	QWeighted calibrated label ranking
RAkEL	Random k-labelsets
RBFN	Radial basis function network
RBM	Restricted Boltzmann machine
RF-PCT	Random forest of predictive clustering trees
ROC	Receiver operating characteristic
ROS	Random over-sampling
RPC	Ranking by pairwise comparison
RUS	Random under sampling
SOM	Self-organizing map
SVD	Single-value decomposition
SVM	Support vector machine
SVN	Support vector network
TF/IDF	Term frequency/inverse document frequency
TN	True negatives
TP	True positives

Chapter 1
Introduction

Abstract This book is focused on multilabel classification and related topics. Multilabel classification is one specific type of classification, classification being one of the usual tasks in the data mining field. Data mining itself can be seen as a step into a broad process, the discovery of new knowledge from databases. The goal of this first chapter is to introduce all these concepts, aiming to set the working context for the topics covered in the following ones. A global outline to this respect is given in Sect. 1.1. Section 1.2 provides an overview of the whole Knowledge Discovery in Databases process. Section 1.3 introduces the essential preprocessing tasks. Then, the different learning styles in use nowadays are explained in Sect. 1.4, and lastly multilabel classification is introduced in comparison with other traditional types of classification in Sect. 1.5.

1.1 Overview

The technological progress in late years has propelled the availability of huge amounts of data. Storage and communication capabilities have grown exponentially, increasing the needs to automatically process all these data. Due to this fact, machine learning techniques have acquired considerable relevance. In particular, the automatic classification of all kind of digital information, including texts, photos, music and videos, is in growing demand. Multilabel classification is the field where methods to perform this task, labeling resources into several non-exclusive categories, are studied and proposed.

This book presents a review of multilabel classification procedures and related techniques, including the analysis of obstacles specifically tied to this class of methods. Experimentation results from the most relevant proposals are also provided. The goal of this first chapter is to set the context multilabel classification belongs to. It starts from the wide view of the whole Knowledge Discovery in Databases (KDD) process, then narrowing the focus until nonstandard classification methods, where multilabel classification is introduced.

© Springer International Publishing Switzerland 2016   1
F. Herrera et al., *Multilabel Classification*,
DOI 10.1007/978-3-319-41111-8_1

1.2   The Knowledge Discovery in Databases Process

The daily activity of millions of users, working and interacting with businesses and institutions, is digitally recorded into databases known as Online Transaction Processing systems. This has led to the availability of huge amounts of data in all kinds of corporations, no matter whether they are small or big companies. Extracting useful knowledge from these data by manual means is extremely difficult, if not impossible. This is why Data Mining (DM) techniques are increasing their popularity as an automatic way of getting the knowledge hidden in the data. This knowledge can be very valuable to support decision-making systems, to describe the structure of the information, to predict future data, and so on.

DM is a very-well-known and solid discipline nowadays, usually seen as one of the steps in the process known as KDD. In [14], KDD is defined as a non-trivial process of identifying valid, novel, potentially useful, and ultimately understandable patterns in data. These patterns would be the result obtained at the end of the process and can be of disparate nature as will be explained below.

Extracting new and useful insights from a database is a process that can be divided into multiple stages. These have been schematically represented in the diagram in Fig. 1.1. The starting point should be understanding the domain the problem belongs to, specifying the goals to achieve. From here, the following steps would be:

1. **Data gathering and integration**: The data needed to accomplish the established objectives can reside in heterogeneous sources, such as relational databases, spreadsheets, comma-separated value files. Once all the data have been gathered,

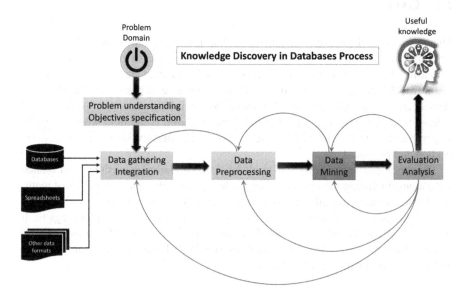

Fig. 1.1 The steps the KDD process is usually divided into

it has to be correctly integrated in a common representation, suitable for next steps in the pipeline.

2. **Data Preprocessing**: Some of the information gathered from data sources can be inconsistent and/or irrelevant. Differences in scales, noise presence, and other anomalies have to be adequately addressed through cleaning methods. By removing unimportant data, only the useful and truly relevant information is selected for further steps. In addition, depending on its nature, several preprocessing methods for data reduction can be applied to data. All of them aim to prepare the data in order to ease the learning conducted in the next phase.

3. **Data Mining**: This is the most known step in the KDD process, and some authors [19] view it as the main stage in the KDD process. Working with the data previously integrated, cleaned, selected and transformed, a DM algorithm has to be chosen to learn from this data. Depending on the objectives set at the beginning, the algorithm can be aimed to group the data samples according to some attributes, learn a model able to automatically classify new samples, etc. In Sect. 1.4 a general overview of DM tasks is provided.

4. **Evaluation and analysis**: The results obtained from the previous step have to be evaluated and analyzed. Interpreting them will assist the user to achieve the desired goals, also easing the overall problem understanding. This would be the useful and non-trivial knowledge extracted from the data.

As the diagram in Fig. 1.1 denotes, all the steps in the KDD process can jump backwards into the pipeline, depending on different conditions. As a consequence, each stage can be applied several times until a certain status is met, aiming to improve data quality in each iteration.

Although DM is considered the essential stage in KDD, most of the effort and time is usually spent in preprocessing tasks [17]. These are responsible of dealing with problems such as missing values, noise presence, outliers, feature and instance selection, and data compression. In the following, a deep analysis of preprocessing and DM tasks is provided.

1.3 Data Preprocessing

Once the data have been retrieved from the source it is stored into, frequently the first step is to prepare it through one or more preprocessing methods. We are generically referring here as *data preprocessing* to several integration, cleaning, transformation, and other data preparation and data reduction tasks. These methods will be run before applying any DM algorithm, easing the process of extracting some useful information.

According to the following statement from [17], data preprocessing duties can be divided into two groups entitled *data preparation* and *data reduction*. Identifying the proper preprocessing to administer to the data can improve the results obtained in subsequent steps of the KDD process. In this section, the main preprocessing tasks are briefly introduced conforming to this grouping criteria.

Data preprocessing includes data preparation, compounded by integration, cleaning, normalization, and transformation of data; and data reduction tasks; such as feature selection, instance selection, and discretization. The result expected after a reliable chaining of data preprocessing tasks is a final dataset, which can be considered correct and useful for further data mining algorithms.

Data Preprocessing in Data Mining, Springer 2015

Data preparation comprehend several assorted tasks, including data cleaning and normalization, dealing with missing values, addressing of noise, and extreme value detection. The main data preparation methods are depicted in Fig. 1.2 (reproduced here with authorization from [17]).

The most usual data normalizations are applied over numeric attributes. The goal is to ease the learning process of DM algorithms, normalizing the scales of values. Normalization is a simple transformation technique whose goal is to adjust the attribute values in order to share a common scale, making them proportional and comparable.

Missing values can be present in a dataset for disparate reasons. They can exist due to a failure in previous data transcription steps, or simply because someone has

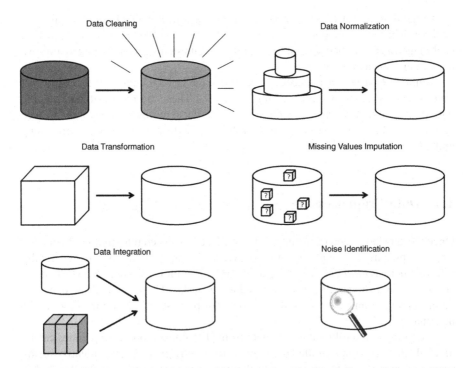

Fig. 1.2 Usual data preparation methods aimed to produce the data before applying any DM algorithm. These would be used just after the data gathering step

left empty a form item. The cleaning can consist in removing the instances with missing values or following some data imputation scheme, usually replacing the missing value by an estimation model.

Noisy data are identified as class noise or attributes containing values which are clearly incorrect, frequently showing random variations that cannot be explained. The techniques to clean noise are similar to the ones aforementioned for missing values.

Building a model from a dataset with a large number of features or instances has a high computational cost. For this reason, data reduction techniques are among the most popular preprocessing methods. These methods can be grouped into the three categories depicted in Fig. 1.3 (reproduced here with authorization from [17]).

Feature selection algorithms aim to reduce the number of features, removing those that are redundant. Reducing the number of features usually helps in simplifying further data mining tasks.

Dimensionality reduction techniques are an alternative to explicit feature selection in some cases, creating a set of artificial features via linear or nonlinear feature combinations. Deep learning methods [24] such as Stacked Autoencoders and Restricted Boltzmann Machines can be also useful for this task.

Fig. 1.3 Visual representation of the three categories that data reduction methods can be grouped into

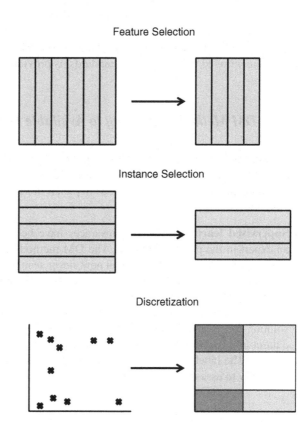

The number of samples in the dataset can be reduced mainly through instance selection techniques [17]. They find the most representative instances in the dataset, selecting them as prototypes prior to applying DM algorithms.

Discretization is a process able to translate a continuous numeric value into a set of adjacent subranges, thus reducing the number of distinct values to consider. The result can be interpreted as a qualitative attribute, instead of a numeric one.

Although the discussed in this section can be considered the most common pre-processing tasks, there are a few more that can be applied when a DM algorithm cannot deal with the data in its current form. Some of them, specific to multilabel classification, will be fully described in Chap. 2.

1.4 Data Mining

The ultimate aim of the data preparatory phases in the KDD process is to apply some DM methods [14], whose duty would be learning from the data to produce useful knowledge. Currently, there is a large collection of proven DM algorithms available. These can be grouped attending to several criteria, depending on the data being labeled or not, the kind of result aimed for, and the model used to represent the knowledge, among others.

In this section, a brief overview of the most remarkable techniques, attending to the aforementioned three grouping standards, is provided.

1.4.1 DM Methods Attending to Available Data

The nature of the available data will affect the kind of DM methods which can be used, also determining the goals that can be set as targets of the learning process. Three main cases can be considered, as depicted in Fig. 1.4. These three categories are as follows:

- **Supervised learning**: The data instances have been previously labeled [9] by an expert in the problem domain. The DM methods can use this information to infer the knowledge needed to label new, never seen before, data instances. In this context, the *label* could be a continuous numerical value or a discrete value. The selected DM method has to work with the data in order to detect relationships between the input attributes, which determine the position of each instance in the solution space, and the target label. Supervised DM methods usually divide the dataset into two (training and test) or three (training, validation and test) disjoint subsets. The label of instances in the test set is not given to the algorithm, being used only to assess its performance.

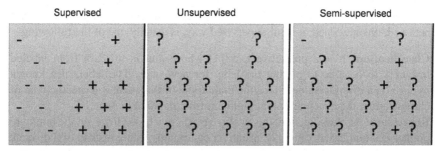

+ Positive | - Negative | ? Unknown class instance

Fig. 1.4 DM Methods attending to the available data nature

- **Unsupervised learning**: All data instances are unlabeled [8], so the DM methods cannot use expert knowledge as foundation of their learning. Since only input attributes are available, there is no point in looking for mappings between inputs and outputs. The values in input attributes are studied to group data instances or search for certain patterns, instead.
- **Semi-supervised learning**: When the dataset contains both labeled and unlabeled samples, DM methods can combine techniques from the two previous categories to accomplish semi-supervised learning tasks [33]. Labeled data instances can be used to induce a model, as in supervised learning, then refining it with the information from unlabeled samples. Analogously, unsupervised tasks can be improved by introducing the clues given by the labeled instances.

1.4.2 DM Methods Attending to Target Objective

From the understanding of the problem domain, starting point in the KDD process, the experts have to specify one or more objectives to achieve. Depending on what kind of target objective is set, a group of DM methods could be applied. The target goals can be grouped into two main categories:

- **Predictive tasks**: The objective is to infer an output value from the information given by the input attributes. The best-known predictive DM tasks are classification and regression. The data samples given to the algorithms have to be labeled, so predictive tasks are usually linked to supervised or semi-supervised learning techniques.
- **Descriptive tasks**: The goal is to extract information about the inner structure of the data samples, represented by the values of their features, or the relationships among them. This way the data can be clustered, patterns can be mined as association rules, anomalies can be detected, and dimensionality can be reduced, among other tasks. These are almost always unsupervised learning tasks, although supervised and semi-supervised learning also have applications in this area.

The individual objective of the learning process, whether it is performed in a supervised, unsupervised or semi-supervised way, is usually one of the following:

- **Classification**: It is a predictive task [1] whose aim is to learn from labeled instances to be able to predict the label for new ones. The label, also known as class, is a discrete value. Since the main topic of this book is a specific type of classification, this is a matter that will be further detailed in Sect. 1.5.
- **Regression**: As classification, this is also a predictive task. However, the output to predict is a numerical value [9], usually a continuous value, instead of a discrete label. The samples the DM method will learn from are annotated with the value associated to each one as output attribute.
- **Clustering**: This is probably the best-known descriptive task [6]. It is usually performed in a unsupervised way, following some kind of similarity metric among the samples features. In addition to its most used application, which is separating the samples in a certain number of groups as similar as possible, clustering is also useful for detecting rare patterns (*outliers*).
- **Pattern mining**: It is also a descriptive DM task. The discovery of unknown relationships among data instances, and between features inside the same instance, is a technique able to extract interesting patterns. These patterns can be expressed as association rules [30], or be used as a way to remove redundant and not informative features, thus reducing the data dimensionality. Supervised techniques can also be used to induct descriptive rules [25] when the patterns have been previously labeled. When the mined data come as a sequence of items, for instance while working with DNA and protein sequences, some specific approaches [13] can be used to obtain useful patterns.
- **Time series forecasting**: Data in time series [22] are compounded by patterns with a chronological relationship. Time series forecasting is a predictive task [12] which has some specific components, such as trends presence, and stationary behavior. These makes it very different from other predictive jobs, such as classification and regression.

1.4.3 DM Methods Attending to Knowledge Representation

Once the task at glance is identified according to the two previous grouping criteria, e.g., the data are labeled and the goal is to classify new instances through supervised learning techniques, a specific representation for the obtained knowledge has to be selected. This will be the last step prior to choosing a particular DM algorithm. Generally, a plethora of algorithms exists for each representation model. The knowledge can be represented as a decision tree, for instance, and there are many distinct algorithms to generate this type of model.

Most models will generate a decision boundary aiming to group the data samples. This boundary can be as simple as a linear one, and as complicated as the ones produced by artificial neural networks in a complicated space. Assuming a plain 2D

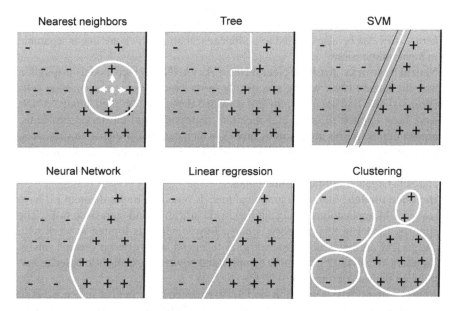

Fig. 1.5 Decision boundaries resulting from some of the best-known learning models

space and samples belonging only to two classes, in Fig. 1.5 the aspect of some of these boundaries has been depicted.

Without aiming to be exhaustive, below some of the most well-studied knowledge representation models are enumerated, along with several of the algorithms associated to them:

- **Instance-based learning**: The most popular algorithm in this category is kNN (*k Nearest Neighbors*) [11]. Depending on the specific algorithm, a bare bones kind of model can be built, e.g., getting the a priori probabilities for each class, or the task can be accomplished without any model. When a new pattern arrives, its nearest neighbors in the search space are obtained, attending to some similarity function, and the class or regression value is predicted from the labels or values present in them.
- **Trees**: Algorithms to induce trees [28] have been applied to classification, regression, and rule discovery, among other DM tasks. A tree recursively divide the search space into smaller groups of instances, following a divide-and-conquer strategy, choosing the best dividing feature in each step according to some quality metric. C4.5 [26] and CART [7] are algorithms used to generate classification trees and regression trees, respectively.
- **Support Vector Machines**: SVMs, also known as SVNs (*Support Vector Network*), are models which maximize the margin between data instances and an hyperplane acting as division boundary [10]. The algorithms aimed to generate this hyperplane are instance-based learners, but the distances are used to adjust the boundary position, instead of getting the most similar neighbors. This model

was originally designed for linear separable problems, but through the so-called *kernel trick* can be also applied to nonlinear scenarios which are linear separable in higher dimensions.

- **Neural Networks**: Artificial Neural Networks (ANN) [27] are models inspired in how the human brain works, specifically in its structure of interconnected cells (*neurons*) and how their outputs are activated depending on their inputs. ANNs have been applied to solve classification, regression, and clustering problems, among others. ANNs are complex models which can follow disparate structures, needing algorithms able to optimize the weights connecting neurons. Some of the possible models are Multilayer Perceptrons, Models, Modular Neural Networks, and Self-Organizing Maps, among others.
- **Bayesian models**: Statistical graphical models based on Bayes theorem [5] have several DM applications, being applied in classification and regression tasks. The simplest model in this category is naïve Bayes, which is limited to work with categorical attributes and assumes total independence between feature values. Bayesian networks [23] are more complex and powerful approaches, based essentially on the same principle.
- **Regression models**: Regression models [9] emerged in the field of statistics, being defined as an iterative procedure to adjust a certain mathematical function to a set of patterns, usually minimizing a precise error metric. They are among the oldest DM models. Some of the most common algorithms generating them are Ordinary least-squares regression, Linear regression, and Logistic regression.
- **Rules**: The research in rule-based systems [15] started more than 50 years ago, being used both for descriptive and predictive tasks. Rule-based classification models are a set of if-then rules, with the features as a conjunction in the precedent and the class label in the consequent. The main advantage of these classifiers is the ease to understand the produced model. Association rules, on the other hand, are aimed to communicate the frequency of patterns in the dataset, as well as the relationship (concurrence, implication, etc.) between them. A set of association rules provides insight about the internal structure of the instances and can be used to describe the data, predict some values from the others presence, group similar samples, etc. The first algorithm aimed to produce this kind of rules was A priori [2]. FP-Growth [20] is also a popular algorithm for association rule mining.

In order to produce the best possible model from the available data, the methods that generate them rely on two foundations, an evaluation metric and an optimization approach. Both will depend on the kind of model to be produced. The evaluation metric can be accuracy for classification models, squared error for regression models, likelihood for Bayesian models, etc. Some common optimization approaches are gradient based methods and evolutionary algorithms, among others.

Table 1.1 Classification problems attending to the output to be predicted

Number of outputs	Output type	Classification kind
1 per instance	Binary	Binary
1 per instance	Multivalued	Multiclass
n per instance	Binary	Multilabel
n per instance	Multivalued	Multidimensional
1 per n instances	Binary/Multivalued	Multiinstance

1.5 Classification

Classification is one of the most popular DM topics. It is a predictive task, usually conducted by means of supervised learning techniques [1]. Classification aims to learn from labeled patterns a model able to predict the label (or class) for future, never seen before, data samples. Some classification algorithms, such as the ones founded on instance-based learning [3], can afford this work without previously building a model.

The set of attributes in a classification dataset is divided into two subsets. The first one contains the input features, the variables that will act as predictors. The second subset holds the output attributes, the so-called class or label assigned to each instance. Classification algorithms induce the model analyzing the correlation between input features and output class. Once a trained model is obtained, it can be used to process the set of features of new data samples getting a class prediction.

Depending on the nature of the second subset of attributes, that containing the class, several kinds of classification problems can be identified. Table 1.1 summarizes the most common configurations, depending on the number of outputs and their types. In the following subsections, each one of them is succinctly described, including some of their applications.

1.5.1 Binary Classification

This is a easiest classification problem we can face [1]. The instances in a binary dataset have only one output attribute, and it can take only two different values. These are usually known as positive and negative, but can also be interpreted as true and false, 1 and 0, or any other combination of two values. A classical example of this task is spam filtering (see Fig. 1.6), in which the classifier learns from the messages' content which ones can be considered as spam.

A binary classifier aims to find a boundary able to separate the instances into two groups, one belonging to the positive class and the other to the negative one. In practice, depending on the input feature space, the distribution of samples can be much more difficult, thus needing additional frontiers between instances. Current

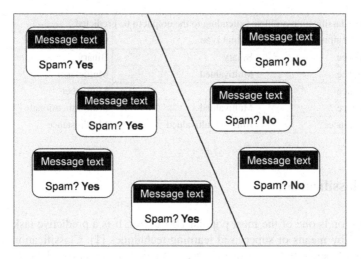

Fig. 1.6 Spam filtering is a typical binary classification problem. The classifier learns from the messages text and predicts if it is spam or not. The problem could be not linear separable, as in this simple example

applications of binary classification are e-mail filtering, to eliminate spam messages, loan analysis, deciding if the customer is economically reliable or not, medical evaluation, determining if a patient has a certain disease or not, and the recognition of all kinds of binary patterns.

1.5.2 Multiclass Classification

A multiclass dataset also has only one output attribute, like in the binary datasets, but it can hold any from a certain set of predefined values. The meaning of each value, and the value itself, would be specific to each application. The set of classes will be finite and discrete, on the contrary the task would not be classification but regression. Class values could have an order relationship or not. One of the best-known multiclass classification examples is iris species identification (see Fig. 1.7). From four characteristics of the flower, i.e., the petal and sepal lengths and widths, the classifier learns how to classify new instances into the corresponding family.

Many multiclass classification algorithms rely on binarization [16], a method that iteratively trains a binary classifier for each class against the others, following a One-vs-All (OVA) approach, or for each pair of classes, using a One-vs-One (OVO) way.

Multiclass classification can be seen as a generalization of binary classification. There is only one output, but it can take any value, while in the binary case it is limited to a subset of two values.

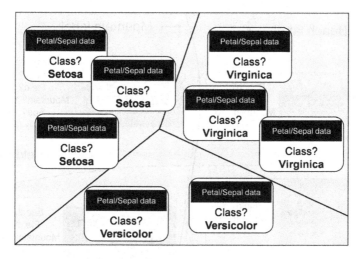

Fig. 1.7 Iris species categorization is a classical multiclass classification problem. The classifier learns from the length and width of petal and sepal, predicting which of the three species the flower belongs to

1.5.3 Multilabel Classification

Unlike the two previous classification models, in multilabel classification [18, 29, 31] each one of the data instances has associated a vector of outputs, instead of only one value. The length of this vector is fixed according to the number of different labels in the dataset. Each element of the vector will be a binary value, indicating if the corresponding label is relevant to the sample or not. Several labels can be active at once. Each distinct combination of labels is known as *labelset*. Figure 1.8 depicts[1] one of the classical multilabel applications, image labeling. The dataset has four labels in total and each picture can be assigned any of them, or even all at once if there were a picture in which the four concepts appear.

Multilabel classification is currently applied in many fields, most of them related to automatic labeling of social media resources such as images, music, video, news, and blog posts. The algorithms used for this task must be able to make several predictions at once, whether it is by transforming the original datasets or by adapting existent binary/multiclass classification algorithms.

---

[1]The pictures used in this example are public domain images taken from the http://www.publicdomainpictures.net website.

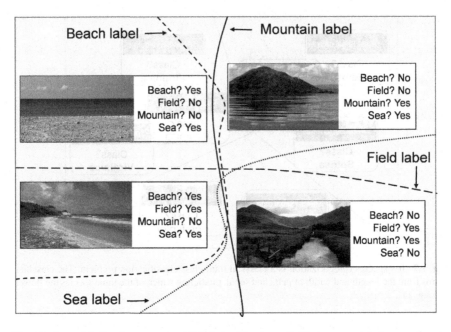

Fig. 1.8 Image labeling is an usual multilabel classification task. Two or more labels are assigned to each image, depending on the elements it contains. There are several overlapping boundaries, one associated to each label

1.5.4   Multidimensional Classification

This type of task can be seen as a generalization of the previous one. Multidimensional datasets [6] also have an output vector associated to each instance, rather than only one value. However, each item in this vector can take any value from a predefined set, not being limited to be binary. Therefore, the relationship between multidimensional and multilabel classification is essentially the same previously mentioned for multiclass and binary classification.

Multidimensional techniques are closely related to that used on the multilabel field, with transformation methods taking a relevant role to face this task. The range of potential applications is also similar. Multidimensional classification is used to categorize images, music, texts and analogous resources, but predicting for each label a value in a larger set than multilabel classification.

1.5.5   Multiple Instance Learning

Unlike all other classification kinds described above, the multiple instance learning paradigm [4], also known as *multiinstance learning*, learns a common class label for a set of input features vectors. It is a very different problem, since each logical data instance is defined not only by a vector of input features, but for a collection of physical instances, each one with a set of input attributes. Each group of instances is usually known as *bag*. The associated class label belongs to the bag, instead of to the individual data instances it holds.

Among the applications of this classification paradigm, one of the most easily understandable is image categorization. The original images are divided into regions or *patches*, so each image is represented by a set of vectors, containing each one the data of a patch. The associated class label could depend on certain objects appearing in some of the patches.

Multiple instance learning can be extended considering multiple class labels as output, rather than only one. From this, casuistic emerges variations such as multiple instance multilabel classification [32]. In the same way, a multiple instance multidimensional classification task could be considered.

> In addition to traditional classification methods, such as the binary and multi-class cases described in this section, there are many other nonstandard classification problems. Multilabel, multidimensional, and multiple instance learning are among them. An enumeration of most of the existent nonstandard classification problems up to now can be found in [21].

References

1. Aggarwal, C.C. (ed.): Data Classification: Algorithms and Applications. CRC Press (2014)
2. Agrawal, R., Srikant, R., et al.: Fast algorithms for mining association rules. In: Proceedings of 20th International Conference on Very Large Data Bases, VLDB'94, pp. 487–499. Morgan Kaufmann (1994)
3. Aha, D.W. (ed.): Lazy Learning. Springer (1997)
4. Amores, J.: Multiple instance classification: review, taxonomy and comparative study. Artif. Intell. **201**, 81–105 (2013)
5. Barber, D.: Bayesian Reasoning and Machine Learning. Cambridge University Press (2012)
6. Bielza, C., Li, G., Larrañaga, P.: Multi-dimensional classification with Bayesian networks. Int. J. Approximate Reasoning **52**(6), 705–727 (2011)
7. Breiman, L., Friedman, J., Stone, C.J., Olshen, R.A.: Classification and Regression Trees. CRC press (1984)
8. Celebi, M.E., Aydin, K. (eds.): Unsupervised Learning Algoritms. Springer (2016)
9. Cherkassky, V., Mulier, F.: Learning from Data: Concepts, Theory and Methods. Wiley-IEEE Press (2007)
10. Cortes, C., Vapnik, V.: Support-vector networks. Mach. Learn. **20**(3), 273–297 (1995)
11. Cover, T., Hart, P.: Nearest neighbor pattern classification. IEEE Trans. Inf. Theory **13**(1), 21–27 (1967)
12. De Gooijer, J.G., Hyndman, R.J.: 25 years of time series forecasting. Int. J. Forecast. **22**(3), 443–473 (2006)
13. Dong, G., Pei, J.: Sequence Data Mining. Springer (2007)

14. Fayyad, U., Piatetsky-Shapiro, G., Smyth, P.: From data mining to knowledge discovery in databases. AI Mag. **17**(3), 37 (1996)
15. Fürnkranz, J., Gamberger, D., Lavrač, N.: Foundations of Rule Learning. Springer (2012)
16. Galar, M., Fernández, A., Barrenechea, E., Bustince, H., Herrera, F.: An overview of ensemble methods for binary classifiers in multi-class problems: experimental study on one-vs-one and one-vs-all schemes. Pattern Recogn. **44**(8), 1761–1776 (2011)
17. García, S., Luengo, J., Herrera, F.: Data Preprocessing in Data Mining. Springer (2015)
18. Gibaja, E., Ventura, S.: A tutorial on multi-label learning. ACM Comput. Surv. **47**(3) (2015)
19. Han, J., Kamber, M., Pei, J.: Data Mining: Concepts and Techniques. Morgan Jaufmann (2011)
20. Han, J., Pei, J., Yin, Y., Mao, R.: Mining frequent patterns without candidate generation: a frequent-pattern tree approach. Data Min. Knowl. Disc. **8**(1), 53–87 (2004)
21. Hernández-González, J., Inza, I., Lozano, J.A.: Weak supervision and other non-standard classification problems: a taxonomy. Pattern Recogn. Lett. **69**, 49–55 (2016)
22. Hyndman, R.J., Athanasopoulos, G.: Forecasting: Principles and practice. OText books (2013)
23. Koller, D., Friedman, N.: Probabilistic Graphical Models. Principles and Techniques. MIT Press (2009)
24. LeCun, Y., Bengio, Y., Hinton, G.: Deep learning. Nature **521**(7553), 436–444 (2015)
25. Novak, P.K., Lavrač, N., Webb, G.I.: Supervised descriptive rule discovery: a unifying survey of contrast set, emerging pattern and subgroup mining. J. Mach. Learn. Res. **10**, 377–403 (2009)
26. Quinlan, J.R.: C4.5: Programs for Machine Learning (1993)
27. Rojas, R.: Neural Networks. A Systematic Study. Springer (1996)
28. Rokach, K., Maimon, O.: Data Mining with Decision Trees, 2nd edn. World Scientific (2015)
29. Tsoumakas, G., Katakis, I., Vlahavas, I.: Mining multi-label data. In: Data Mining and Knowledge Discovery Handbook, pp. 667–685. Springer (2010)
30. Zhang, C., Zhang, S.: Association Rule Mining. Springer (2002)
31. Zhang, M.L., Zhou, Z.H.: A review on multi-label learning algorithms. IEEE Trans. Knowl. Data Eng. **26**(8), 1819–1837 (2014)
32. Zhou, Z.H., Zhang, M.L., Huang, S.J., Li, Y.F.: Multi-instance multi-label learning. Artif. Intell. **176**(1), 2291–2320 (2012)
33. Zhu, X., Goldberg, A.B.: Introduction to Semi-supervised Learning. Morgan & Claypool Publishers (2009)

Chapter 2
Multilabel Classification

Abstract This book is concerned with the classification of multilabeled data and other tasks related to that subject. The goal of this chapter is to formally introduce the problem, as well as to give a broad overview of its main application fields and how it have been tackled by experts. A general introduction to the matter is provided in Sect. 2.1, followed by a formal definition of the multilabel classification problem in Sect. 2.2. Some of the main application fields of multilabel classification are portrayed in Sect. 2.3. Lastly, the approaches followed to face this duty are introduced in Sect. 2.4.

2.1 Introduction

Multilabel classification is a predictive data mining task with multiple real-world applications, including the automatic labeling of many resources such as texts, images, music, and video. The learning from multilabel data can be accomplished through different approaches, such as data transformation, method adaptation, and the use of ensembles of classifiers.

This chapter begins by formally defining the multilabel classification problem, introducing the mathematical notation and terminology that will be used throughout this book. Then, the different areas in which multilabel classification is applied nowadays will be outlined, and the repositories this kind of data can be obtained from are introduced.

The learning from multilabel data is being currently faced through disparate approaches, including data transformation and adaptation of traditional classification methods. The use of ensembles of classifiers is also quite popular in this field. In addition, some specific aspects, such as the use of label dependency information, the problems of high dimensionality, and label imbalance, must be considered. All these topics will be further described, along with an enumeration of the main multilabel software tools currently available.

© Springer International Publishing Switzerland 2016   17
F. Herrera et al., *Multilabel Classification*,
DOI 10.1007/978-3-319-41111-8_2

2.2   Problem Formal Definition

The main difference between traditional[1] and multilabel classification is in the output expected from trained models. Where a traditional classifier will return only one value, a multilabel one has to produce a vector of output values. Multilabel classification can be formally defined as follows.

2.2.1   Definitions

Definition 2.1 Let $\mathcal{X}$ denote the input space, with data samples $X \in A_1 \times A_2 \times ... \times A_f$, being f the number of input attributes and $A_1, A_2, ..., A_f$ arbitrary sets. Therefore, each instance X will be obtained as the cartesian product of these sets.

Definition 2.2 Let $\mathcal{L}$ be the set of all possible labels. $\mathcal{P}(\mathcal{L})$ denotes the powerset of $\mathcal{L}$, containing all the possible combinations of labels $l \in \mathcal{L}$ including the empty set and $\mathcal{L}$ itself. $k = |\mathcal{L}|$ is the total number of labels in $\mathcal{L}$.

Definition 2.3 Let $\mathcal{Y}$ be the output space, with all the possible vectors $Y, Y \in \mathcal{P}(\mathcal{L})$. The length of Y always will be k.

Definition 2.4 Let $\mathcal{D}$ denote a multilabel dataset, containing a finite subset of $A_1 \times A_2 \times ... \times A_f \times \mathcal{P}(\mathcal{L})$. Each element $(X, Y) \in \mathcal{D} | X \in A_1 \times A_2 \times ... \times A_f, Y \in \mathcal{P}(\mathcal{L})$ will be an instance or data sample. $n = |\mathcal{D}|$ will be the number of elements in $\mathcal{D}$.

Definition 2.5 Let $\mathcal{F}$ be a multilabel classifier, defined as $\mathcal{F} : \mathcal{X} \to \mathcal{Y}$. The input to $\mathcal{F}$ will be any instance $X \in \mathcal{X}$, and the output will be the prediction $Z \in \mathcal{Y}$. Therefore, the prediction of the vector of labels associated with any instance can be obtained as $Z = \mathcal{F}(X)$.

2.2.2   Symbols

From these definitions, the following list of symbols, which will be used in this chapter and the next ones, is derived:

$\mathcal{D}$ Any multilabel dataset (MLD).
n The number of data instances in $\mathcal{D}$.
$\mathcal{L}$ The full set of labels appearing in $\mathcal{D}$.
l Any of the labels in $\mathcal{L}$.
k Total number of elements in $\mathcal{L}$.
X The set of input attributes of any instance.

---

[1]In general, we will refer to binary and multiclass classification, which are the most well-known classification kinds, as *traditional* classification.

f The number of attributes comprising X.
$\mathcal{X}$ The full input space in $\mathcal{D}$, consisting of all X instances.
Y The set of output labels (*labelset*) of any instance.
$\mathcal{Y}$ The full output space in $\mathcal{D}$, comprised by all Y instances.
Z The labelset predicted by the classifier.

While referring to specific individual instances, the usual notation $\mathcal{D}_i$ will be used. Analogously, Y_i will make reference to the true labelset of instance i, Z_i the labelset predicted by the classifier to the ith instance, and so on.

The metrics aimed to evaluate the performance of a classifier (they will be described in the next chapter) compute the differences between Y_i and Z_i, that is, the real labelset associated with each instance and the one predicted by the classifier. The goal while training a multilabel classifier would be reducing the error ratio brought by these metrics.

2.2.3 Terminology

In addition to the previous notation, the following terms will be frequently used throughout the text:

- **MLD/MLDs**: Any multilabel dataset or group of multilabel datasets.
- **MLC**: Any multilabel classifier or multilabel classification algorithm.
- **Instance/sample**: A row in an MLD, including its input attributes and associated labelset.
- **Attributes/features**: Generally used to refer to the set of input attributes in the MLD, without including the output labelset.
- **Label**: Any of the output attributes associated with an instance.
- **Labelset**: A vector of labels $\{0, 1\}^k$ whose length will be always k.
- **True labelset**: The labelset a sample in the MLD is annotated with.
- **Predicted labelset**: The labelset an MLC is giving as output for a new sample.

2.3 Applications of Multilabel Classification

Once the main concepts linked to multilabel classification have been introduced, the next question arising possibly is what can it be used for. As has been explained in the previous section, a multilabel classifier aims to predict a set of relevant labels for a new data instance.

In this section, several application areas that can take advantage of this functionality are portrayed. In the following chapter, specific use cases belonging to each one of these areas will be detailed.

2.3.1   Text Categorization

Multilabel classification has its roots as a solution for organizing text documents into several not mutually exclusive categories. This is why there are so many publications regarding this topic [15, 21, 40–42]. Most of them will be further described in the next chapter.

Textual documents can be found anywhere, from big companies which store all kind of agreements and reports to individuals filing their invoices and electronic mail messages. All published books and magazines, our historic medical records, as well as articles in electronic media, blog posts, question–answering forum messages, etc., are text documents also. Most of them can be classified into more than one category, thus the usefulness of multilabel classification to accomplish this kind of work.

The usual process to transform a set of text documents into an MLD relies on text mining techniques. The source documents are parsed, uninformative words are removed, and vectors with each word occurrence among the documents are computed. At the end of this process, each document is described by a row in the MLD, and the columns correspond to words and their frequencies or some other kind of representation such as TF/IDF (*Term Frequency/Inverse Document Frequency*, [47]).

2.3.2   Labeling of Multimedia Resources

Although documents containing only text were the most frequent ones in the past, nowadays images, videos, and music are commonly used resources due to the huge growth experienced by storage and communication technologies. Attending to 2015 YouTube statistics, an estimated 432 000 new videos are uploaded every day. The number of new music and sound clips published each day is also impressive, and new images and photographs posted everywhere, from blogs to Web sites such as Tumblr, reach the millions per day barrier.

Multilabel classification has been used to label all these types of resources [5, 7, 28, 48, 56, 58], identifying the objects which appear in sets of images, the emotions produced by music clips, or the concepts which can be derived from video snips. This way, the huge number of new resources can be correctly classified without needing human intervention, something that would be rather expensive.

Even though an image can be represented as a vector of color values, taking each pixel as a column, usually they are preprocessed in order to obtain the most relevant features. For doing so, segmentation techniques aimed at extracting boundaries are applied, the image can be convoluted to transform the pixel space into an energy space, etc. Similar procedures are put in practice with other multimedia resources.

2.3.3 Genetics/Biology

Proteomics and genomics are research fields which have experienced a huge growth in late years. As a result, immense databases with protein sequences have been produced, but only a small fraction of them have been studied to determine their function. Analyzing protein properties and gene expression is a very costly task, but DM techniques can accelerate the process and make it cheaper.

The application of multilabel classification in this area [25, 29] aimed at predicting the biologic functions of genes. Each gene can express more than one function at once, hence the interest in using MLC techniques to face this problem. The features used as predictors are usually the protein's *motifs*, which are traits about its internal structure. Structural motifs indicate how the elements in a secondary structural layer are connected. Certain patterns, such as short chains of amino acids, can be identified and used as predictive features.

2.3.4 Other Application Fields

In addition to the ones referenced in the previous sections, multilabel classification has been utilized in many other applications, both with public and with private datasets, sometimes ad hoc generated for a specific need. Most of them could be included in the first two categories, since eventually the goal is to categorize texts, sounds, images, and videos. The following are among the most interesting ones:

- The analysis of nonverbal expressions in speech is the focus in [49], aiming at detecting how people feel. The goal is to predict several not mutually exclusive affective states at once. The predictive attributes are a set of vocal features such as intonation and speech rate.
- In [19], the authors propose a system to automatically classify patent records. The addressed problem is easing the search of previous documents according to inventive principles. In the end, this proposal can be seen as another text categorization solution.
- The method presented in [50] aims to improve the process of searching for relevant information in Twitter. Five different labels are defined to classify twits, including news, opinions, and events. Several of them can be assigned simultaneously to the same twit.
- Analyzing complex motion in events is the goal of the system proposed in [17]. It combines trajectory and multilabel hypergraphs of moving targets in video sequences, detecting the relationship between the low-level features and high-level concepts which are to be predicted.

In general, multilabel classification could be suitable for any scenario in which some kind of information, no matter its type as long as it can be processed by an MLC algorithm, is assigned to two or more categories simultaneously.

2.3.5 MLDs Repositories

Usually, when new datasets are generated, the authors of each original work publicly provide the MLDs they have produced, so that other researchers can use them in their own studies. Notwithstanding, most MLDs can be obtained directly from some of the following data repositories:

- **MULAN**: The well-known multilabel learning package [55] has an associated data repository [54]. As of 2016, more than 20 MLDs are available in it. These MLDs are in ARFF[2] (*Attribute-Relation File Format*), and each one has an associated XML file describing the labels and their relationships. The XML file is needed, since the position of the attributes acting as labels is not assumed to be at the end.
- **KEEL**: KEEL [2] is a open source software providing lots of preprocessing and DM algorithms. It has an associated data repository [1] with almost 20 MLDs. Along with the full datasets, 10-fcv and 5-fcv partitions are also supplied. The KEEL file format, based on ARFF, includes the name of output attributes in the header; therefore, a separate XML file is not needed.
- **MEKA**: MEKA is a multilabel software developed on top of WEKA [37]. As the previous ones, it has also associated a data repository [45] with over 20 MLDs. The file format is also ARFF based, and the information needed to know which attributes are the labels, specifically the number of labels, is included in the header. Some multidimensional datasets are included.
- **RUMDR**: *R Ultimate Multilabel Dataset Repository* [11] provides a compilation of all MLDs publicly available, as well as an R package which eases the partitioning and exporting to several formats, including MULAN, KEEL, MEKA, LibSVM, and CSV.

A common group of MLDs is available in any of the aforementioned repositories, but in the MULAN and MEKA repositories, some specific ones, not available in the other sites, can be found. The partitions used in the experiments conducted in following chapters can be downloaded from the GitHub repository associated with the book [12].

2.4 Learning from Multilabel Data

The process to obtain a multilabel classification model is similar to that used for traditional classification, usually following supervised learning techniques. Most algorithms rely on an initial training phase. It depends on previously labeled samples to adjust the parameters of the model. Once trained, the model can be used to predict the labelset for new instances.

---

[2]ARFF is the file format used by WEKA [37].

When it comes to learning from multilabel data, two main approaches have been applied: data transformation and method adaptation. The former is based on transformation techniques which, applied to the original MLDs, are able to produce one or more binary or multiclass datasets. These can be processed with traditional classifiers. The latter aims for adapting existent classification algorithms, so they can deal with multilabel data, producing several outputs instead of only one. A third alternative, which naturally emerges from the first one, is the use of ensembles of classifiers.

A topic closely related to MLC is label ranking [57], whose goal is to map each instance in a dataset to a set of labels, establishing a total order relationship among them. That is, the output of these algorithms is a ranking of labels, assigning to each one a certain weight. A multilabel classifier can use this label ranking to decide which labels are relevant to the instance, applying a cut threshold that has to be computed in some way. In [8, 30, 38], different proposals on how to do multilabel classification founded on label rankings are presented.

This section introduces the three main approaches to MLC: data transformation, method adaptation, and ensembles of classifiers. In addition, it also advances some of the aspects most often alluded to in multilabel learning, as is taking advantage of label correlation information, the high-dimensionality problem, and the learning from imbalanced data task.

2.4.1 The Data Transformation Approach

MLC is a harder task than traditional classification, since the classifier has to predict several outputs at once. One of the first approaches to arise for solving this problem was to transform it, producing one or more simpler problems. The transformation idea is all about to get an MLD and generate datasets that can be processed by binary or multiclass classifiers. Commonly, the output produced by those classifiers has to be backtransformed in order to obtain the multilabel prediction.

Some of the simplest methods originally proposed in the literature are the ones described below:

- **Selecting a single label**: It is the transformation named *Model-s* in [5], *s* standing for *single label*. When a sample is associated with a set of labels, one of them is chosen as single class. This selection can be random or be based on some heuristic method. The result produced by this transformation is a multiclass dataset having the same number of instances than the original one, but each sample has only one class.
- **Ignoring multilabel instances**: Dismissing all the samples associated with more than one label, this transformation obtains a new dataset of multiclass nature, with only one label per instance. It is introduced as *Model-i* in [5], *i* standing for *ignore*. The resulting dataset usually will have less instances than the original one, unless none of the samples had two or more labels. Since there is only one label per instance, any multiclass classifier can be used.

- **Unfolding samples with multiple labels**: Introduced in [53] as PT5 (*Problem Transformation 5*), this method decomposes each instance into as many instances as labels it contains, cloning the input attributes and assigning to each sample one of the labels. A weight can be optionally assigned to each label, depending on its distribution on the MLD. The output of this transformation is also a multiclass dataset, in this case containing more samples than the original one.
- **Using the labelset as class identifier**: Instead of discarding labels or samples, the method presented in [5] as *Model-n*, the *n* standing for *new class*, proposes using each different combination of labels (each labelset) as identifier of a new class. The resulting dataset has the same number of instances, but only one class. Therefore, it can be processed with any multiclass classifier. Nowadays, this transformation method is best known as LP (*Label PowerSet*).
- **Applying binarization techniques**: Binarization techniques were already used to deal with multiclass classification using binary classifiers; thus, they were a clear choice also for multilabel classification. The most usual approach, used in [35], consists in training k classifiers, each for one label, taking the instances in which the labels appear as positive and all the others as negative. Another proposal, called *cross-training* in [5], also trains k classifiers but prefers to use samples with multiple labels always as positive. The former way is currently known as BR (*Binary Relevance*), and it is possibly the most common transformation method.

For each specific transformation, a method to build the predicted labelset has to be defined. For instance, if the BR transformation is applied to the original MLD, the individual predictions obtained from the binary classifiers have to be joined making up the corresponding labelset.

In late years, the LP and BR transformations have been the foundation for multiple MLC solutions, including many of which are based on ensembles of binary and multiclass classifiers. Additional details related to these techniques will be provided in Chap. 4, along with some experimental results.

2.4.2   The Method Adaptation Approach

Automatic classification [26] has been a traditional DM task for years. Throughout this time, several families of algorithms [18] have been gradually developed, tested, and fine-tuned. These algorithms are an essential foundation to develop new, more specific ones, including those aimed at working with MLDs.

Some classification models were initially designed to solve binary problems and then extended to also consider multiclass cases [4, 31]. An example of this is SVM [9]. By contrast, other approaches are able to deal with several classes with great simplicity. A kNN classifier [20], for instance, can predict the most frequent class among the neighbors of the new instance, whether there are two or more classes. Analogously, trees, classification rules, neural networks, and statistical models, among others, have been used to tackle both binary and multiclass classifications.

The lessons learned from adapting binary classifiers to the multiclass scenario, such as binarization, voting methods, and divide-and-conquer procedures, have been also useful while adapting proven algorithms to tackle multilabel classification. The main difficulty is to build a model capable of predicting several outputs at once. Again, some approaches are easily adaptable, whereas others require more effort. A kNN-based algorithm can take the multiple outputs present on its neighbors to elaborate its own prediction, but a SVM has to find the maximum margin boundaries between multiple labels.

Proposals of new multilabel classifiers based on the adaptation approach have proliferated lately [34]. Many of them will be introduced in Chap. 5, and the most representative ones will be detailed and experimentally tested.

2.4.3 Ensembles of Classifiers

Classification ensemble is a widespread technique aimed to improve the performance obtained by individual classifiers. An ensemble is compounded by a set of classifiers, whose output is usually combined by weighted or unweighted averaging [24, 46]. The point is that a group of weak classifiers, with different biases, is able to perform better than a strong classifier, following the famous Wolpert's No Free Lunch Theorem [59]. One key aspect in ensemble construction is diversity. The more diverse the individual classifiers are, the more likely they have different biases. Diversity can be achieved through training homogeneous models with different data, for instance with bagging techniques [6], or alternatively by using heterogeneous classification models.

Ensembles of binary classifiers have been used to face multiclass classification [31], either by way of OVA or by way of OVO decompositions. Furthermore, ensembles are usually applied to fight with problems such as imbalance [32]. Therefore, it is not surprising that ensemble techniques are also applied in the multilabel field. In fact, it is the most usual approach to do multilabel classification.

As will be shown in Chap. 6, a plethora of ensembles, made up of binary and multiclass classifiers, have been defined in the literature as a way to fulfill the multilabel classification needs. Some of the best-known proposals are ensembles of classifier chains [44] and ensembles of pruned sets [43]. Both will be further detailed along with other related ideas.

2.4.4 Label Correlation Information

Many of the published multilabel classification algorithms rely on a process to simplify the original problem, producing several easier to confront ones. Through this
process, usually a complete independence between the labels is assumed. However,
most researchers highlight [3, 22, 52, 60] the importance to take into account label
dependency information. Allegedly, these correlations would help in designing better
classification models.

The BR transformation method, as well as many algorithms based on it, builds an
individual classifier for each label. The classifiers are completely independent, so the
prediction made by one of them does not influence how the others make their work.
This is a first approach to label dependency, which consists in assuming that they
are fully independent. However, there are use cases where the presence of a certain
label could determine whether another one is also more likely to be present or not.
In scene labeling, for instance, the probability of the label beach would be higher
if the label sea is also relevant.

A second mechanism is to implicitly incorporate label dependency information
into the classification process. The LP transformation, and some other algorithms
based on taking full or partial labelsets as class identifiers, follows this way. Since sets
of labels are treated like a unit, the dependency among them is implicitly embedded
into the model through the training process. A similar approach, but relying on binary
classifiers instead of multiclass ones, is the one based on chains of classifiers [44].
This technique introduces the label predicted by one classifier into the data given as
input to the next one, as will be detailed in Chap. 6.

Explicit procedures for taking advantage of label correlation information have
been also developed. The authors of the CML (*Collectible Multilabel*) algorithm
[33], for instance, propose the use of conditional random fields to model correlations
between label pairs. The authors of [36] define conditional dependency networks
to capture correlations among labels. Similar statistical models are being used to
explicitly represent this information.

2.4.5 High Dimensionality

High dimensionality is a problem which affects multilabel classification even more
than it does in traditional classification. Usually, three dimensionality spaces are
considered:

- **Instance space**: There are MLDs with millions of instances, but this is something
  increasingly common in all contexts, including traditional classification, due to
  the rise of big data.
- **Input feature space**: Although there are binary and multiclass datasets with many
  input features, MLDs stand out in this area as they usually have dozens of thousands

of features. The dimensionality of this space is what mostly makes difficult the learning process.

- **Output attribute space**: Before multilabel classification was considered, the classifiers only had to deal with one output attribute. Whether it was binary or multivalued, there was only one output, always. By contrast, all MLDs have several outputs and many of them have thousands, one for each label. Due to this fact, the dimensionality treatment in MLDs is a more complex topic than in traditional classification.

That learning from a high-dimensional input space (i.e., a large number of features) imposes serious difficulties is something well known. Even there is an ad hoc expression to name this problem, *the curse of dimensionality*. Therefore, it is an obstacle deeply studied and analyzed in DM, with dozens of different proposals to face it to some extent [51].

There are non-supervised methods, such as Principal Component Analysis (PCA) [39] and Latent Semantic Indexing (LSI) [27], able to reduce dimensionality by analyzing only the input features. These can be applied to MLDs, since no knowledge about the output variables is needed. The supervised approaches, characterized for using the output attributes to infer which input features provide more useful information, have to be adapted prior to be used in the multilabel context.

Dimensionality in the label space is the least studied problem until now, with only a few proposals. Most of them will be detailed in Chap. 7, specifically devoted to this topic.

2.4.6 Label Imbalance

The learning from imbalanced data is another of the casuistics intrinsically linked to multilabel classification. Like high dimensionality in the input feature space, imbalance is also a well-studied problem in traditional classification. There are many proven techniques to face imbalance in binary classification, and many of them have been adapted to work with multiclass datasets. A few of these techniques will be described in Chap. 8, including how some of them have been adjusted to deal with MLDs.

The differences among label distributions in MLDs arise firstly by their own nature. In an MLD with thousands of labels, for instance categorizing text documents, it is usual that some of them are much more frequent than others. More specialized documents would have rare labels, which will be in minority against the most common ones.

Secondly, the transformations applied to multilabel data tend to increase the imbalance between labels. This is the case of BR. Taking as positive only the instances in which a certain label appears and as negative all other samples, the original distribution changes affecting the representation of the considered label. The situation is even worse if this label is already a minority label.

> Imbalance in multilabel classification is one of the specificities more recently tackled in the literature. Specific metrics [13] and several methods to balance label distributions [14, 16, 23] have been proposed. Most of them will be presented in Chap. 8.

2.5 Multilabel Data Tools

To conclude this chapter, in which the topics of what multilabel classification is, what it is used for, and how it has been faced have been dealt, the main tools currently used to work with this kind of information are briefly introduced. They will be explained in more detail in Chap. 9.

Although multilabel data can be analyzed and multilabel classification conducted with any ad hoc software, specifically designed to accomplish a certain task, there are also available some generic tools. Many researchers and practitioners have relied on them in late years. The most noteworthy are the following:

- **MULAN**: Presented in 2010 [55], it is the most mature multilabel software and probably the most widely used. MULAN is an open source library written in Java. It provides a programming interface (API) to help in the development of multilabel classification applications, including also evaluation of results and label ranking generation. Several of the most common transformations and MLC algorithms are already included, ready to use. Unlike WEKA [37], the tool it is based on, MULAN does not offer a graphic user interface (GUI) to ease the accomplishment of these tasks.
- **MEKA**: As MULAN, MEKA [45] is also founded on WEKA, but unlike MULAN it supplies a GUI from which the user can load MLDs, perform some exploration tasks, and apply different MLC algorithms. Since the GUI is very similar to that of WEKA, WEKA users will immediately feel comfortable with this tool. It is also open source and has been developed in Java.
- **mldr**: R is one of the most used tools to explore and analyze data nowadays, but it lacked the ability to present the specificities of MLDs. The mldr [10] package adds this functionality to R, providing functions to load, modify, and write MLDs, as well as general exploratory analysis methods. The package also includes a Web-based GUI from which most tasks can be easily performed.
- **mldr.datasets**: Built upon the infrastructure provided by the preceding one, this R package [11] eases the process of obtaining multilabel datasets. In addition, it provides methods aimed to retrieve basic information about them, including citation data, as well as to facilitate the partitioning process.

- **scikit-multilearn**: Based on the well-known scikit-learn Python module, scikit-multilearn[3] is an extension still in early development. The current version is 0.0.1 and it provides BR, LP, and RAkEL implementations, with many other methods whose development is in progress.

As mentioned above, the details about how to use these tools to explore MLDs and conduct experiments with them are the main topic of Chap. 9.

References

1. Alcala-Fdez, J., Fernández, A., Luengo, J., Derrac, J., García, S., Sánchez, L., Herrera, F.: KEEL multi-label dataset repository. http://sci2s.ugr.es/keel/multilabel.php
2. Alcala-Fdez, J., Fernández, A., Luengo, J., Derrac, J., García, S., Sánchez, L., Herrera, F.: KEEL data-mining software tool: data set repository and integration of algorithms and experimental analysis framework. J. Mult-valued Log. Soft Comput. **17**(2–3), 255–287 (2011)
3. Alvares-Cherman, E., Metz, J., Monard, M.C.: Incorporating label dependency into the binary relevance framework for multi-label classification. Expert Syst. Appl. **39**(2), 1647–1655 (2012)
4. Aly, M.: Survey on multiclass classification methods. In: Technical Report, pp. 1–9. California Institute of Technology (2005)
5. Boutell, M., Luo, J., Shen, X., Brown, C.: Learning multi-label scene classification. Pattern Recogn. **37**(9), 1757–1771 (2004)
6. Breiman, L.: Bagging predictors. Mach. Learn. **24**(2), 123–140 (1996)
7. Briggs, F., Lakshminarayanan, B., Neal, L., Fern, X.Z., Raich, R., Hadley, S.J.K., Hadley, A.S., Betts, M.G.: Acoustic classification of multiple simultaneous bird species: a multi-instance multi-label approach. J. Acoust. Soc. Am. **131**(6), 4640–4650 (2012)
8. Brinker, K., Hüllermeier, E.: Case-based multilabel ranking. In: Proceedings of 20th International Joint Conference on Artificial Intelligence, IJCAI'07, pp. 702–707. Morgan Kaufmann (2007)
9. Burges, C.J.C.: A tutorial on support vector machines for pattern recognition. Data Min. Knowl. Disc. **2**(2), 121–167 (1998)
10. Charte, F., Charte, D.: Working with multilabel datasets in R: the mldr package. R J. **7**(2), 149–162 (2015)
11. Charte, F., Charte, D., Rivera, A.J., del Jesus, M.J., Herrera, F.: R Ultimate multilabel dataset repository. In: Proceedings of 11th International Conference on Hybrid Artificial Intelligent Systems, HAIS'16, vol. 9648, pp. 487–499. Springer (2016)
12. Charte, F., Rivera, A.J., del Jesus, M.J., Herrera, F.: Multilabel classification. In: Problem Analysis, Metrics and Techniques Book Repository. https://github.com/fcharte/SM-MLC
13. Charte, F., Rivera, A.J., del Jesus, M.J., Herrera, F.: Addressing imbalance in multilabel classification: measures and random resampling algorithms. Neurocomputing **163**, 3–16 (2015)
14. Charte, F., Rivera, A.J., del Jesus, M.J., Herrera, F.: MLSMOTE: approaching imbalanced multilabel learning through synthetic instance generation. Knowl.-Based Syst. **89**, 385–397 (2015)
15. Charte, F., Rivera, A.J., del Jesus, M.J., Herrera, F.: QUINTA: a question tagging assistant to improve the answering ratio in electronic forums. In: Proceedings of IEEE International Conference on Computer as a Tool, EUROCON'15, pp. 1–6. IEEE (2015)
16. Chen, K., Lu, B., Kwok, J.: Efficient classification of multi-label and imbalanced data using min-max modular classifiers. In: Proceedings of IEEE International Joint Conference on Neural Networks, IJCNN'06, pp. 1770–1775 (2006)

---

[3]http://scikit.ml.

17. Chen, X., Zhan, Y., Ke, J., Chen, X.: Complex video event detection via pairwise fusion of trajectory and multi-label hypergraphs. Multimedia Tools Appl. 1–22 (2015)
18. Cherkassky, V., Mulier, F.: Learning from Data: Concepts. Theory and Methods. Wiley-IEEE Press (2007)
19. Cong, H., Tong, L.H.: Grouping of triz inventive principles to facilitate automatic patent classification. Expert Syst. Appl. **34**(1), 788–795 (2008)
20. Cover, T., Hart, P.: Nearest neighbor pattern classification. IEEE Trans. Inf. Theory **13**(1), 21–27 (1967)
21. Crammer, K., Dredze, M., Ganchev, K., Talukdar, P.P., Carroll, S.: Automatic code assignment to medical text. In: Proceedings of Workshop on Biological, Translational, and Clinical Language Processing, BioNLP'07, pp. 129–136. Association for Computational Linguistics (2007)
22. Dembszynski, K., Waegeman, W., Cheng, W., Hüllermeier, E.: On label dependence in multi-label classification. In: ICML Workshop on Learning from Multi-label data, pp. 5–12 (2010)
23. Dendamrongvit, S., Kubat, M.: Undersampling approach for imbalanced training sets and induction from multi-label text-categorization domains. In: New Frontiers in Applied Data Mining, *LNCS*, vol. 5669, pp. 40–52. Springer (2010)
24. Dietterich, T.: Ensemble methods in machine learning. In: Multiple Classifier Systems. *LNCS*, vol. 1857, pp. 1–15. Springer (2000)
25. Diplaris, S., Tsoumakas, G., Mitkas, P., Vlahavas, I.: Protein classification with multiple algorithms. In: Proc. 10th Panhellenic Conference on Informatics, PCI'05, vol. 3746, pp. 448–456. Springer (2005)
26. Duda, R., Hart, P., Stork, D.: Pattern Classification, 2nd edn. John Wiley (2000)
27. Dumais, S., Furnas, G., Landauer, T., Deerwester, S., Deerwester, S., et al.: Latent semantic indexing. In: Proceedings of 4th Text Retrieval Conference, TREC-4, pp. 105–115. NIST (1995)
28. Duygulu, P., Barnard, K., de Freitas, J., Forsyth, D.: Object recognition as machine translation: learning a lexicon for a fixed image vocabulary. In: Proceedings of 7th European Conference on Computer Vision, ECCV'02, vol. 2353, pp. 97–112. Springer (2002)
29. Elisseeff, A., Weston, J.: A kernel method for multi-labelled classification. In: Advances in Neural Information Processing Systems, vol. 14, pp. 681–687. MIT Press (2001)
30. Fürnkranz, J., Hüllermeier, E., Loza Mencía, E., Brinker, K.: Multilabel classification via calibrated label ranking. Mach. Learn. **73**, 133–153 (2008)
31. Galar, M., Fernández, A., Barrenechea, E., Bustince, H., Herrera, F.: An overview of ensemble methods for binary classifiers in multi-class problems: experimental study on one-vs-one and one-vs-all schemes. Pattern Recogn. **44**(8), 1761–1776 (2011)
32. Galar, M., Fernández, A., Barrenechea, E., Bustince, H., Herrera, F.: A review on ensembles for the class imbalance problem: bagging, boosting, and hybrid-based approaches. IEEE Trans. Syst. Man Cybern. Part C Appl. Rev. **42**(4), 463–484 (2012)
33. Ghamrawi, N., McCallum, A.: Collective multi-label classification. In: Proceedings of 14th ACM International Conference on Information and Knowledge Management, CIKM'05, pp. 195–200. ACM (2005)
34. Gibaja, E., Ventura, S.: A tutorial on multi-label learning. ACM Comput. Surv. **47**(3) (2015)
35. Gonçalves, T., Quaresma, P.: A preliminary approach to the multilabel classification problem of Portuguese juridical documents. In: Proceedings of 11th Portuguese Conference on Artificial Intelligence, EPIA'03, pp. 435–444. Springer (2003)
36. Guo, Y., Gu, S.: Multi-label classification using conditional dependency networks. In: Proceedings of 22th International Joint Conference on Artificial Intelligence, *IJCAI'11*, vol. 2, pp. 1300–1305 (2011)
37. Holmes, G., Donkin, A., Witten, I.H.: WEKA: a machine learning workbench. In: Proceedings of 2nd Australian and New Zealand Conference on Intelligent Information Systems, ANZIIS'02, pp. 357–361 (2002)
38. Hüllermeier, E., Fürnkranz, J., Cheng, W., Brinker, K.: Label ranking by learning pairwise preferences. Artif. Intell. **172**(16), 1897–1916 (2008)

39. Jolliffe, I.: Introduction. In: Principal Component Analysis, pp. 1–7. Springer (1986)
40. Katakis, I., Tsoumakas, G., Vlahavas, I.: Multilabel text classification for automated tag suggestion. In: Proceedings of European Conference on Machine Learning and Principles and Practice of Knowledge Discovery in Databases, ECML PKDD'08, pp. 75–83 (2008)
41. Klimt, B., Yang, Y.: The enron corpus: a new dataset for email classification research. In: Proc eedings of 15th European Conference on Machine Learning, ECML'04, pp. 217–226. Springer (2004)
42. Lewis, D.D., Yang, Y., Rose, T.G., Li, F.: RCV1: a new benchmark collection for text categorization research. J. Mach. Learn. Res. **5**, 361–397 (2004)
43. Read, J., Pfahringer, B., Holmes, G.: Multi-label classification using ensembles of pruned sets. In: Proc eedings of 8th IEEE International Conference on Data Mining, ICDM'08, pp. 995–1000. IEEE (2008)
44. Read, J., Pfahringer, B., Holmes, G., Frank, E.: Classifier chains for multi-label classification. Mach. Learn. **85**, 333–359 (2011)
45. Read, J., Reutemann, P.: MEKA multi-label dataset repository. http://sourceforge.net/projects/meka/files/Datasets/
46. Rokach, L.: Pattern classification using ensemble methods. World Scientific (2009)
47. Salton, G., Fox, E.A., Wu, H.: Extended Boolean information retrieval. Commun. ACM **26**(11), 1022–1036 (1983)
48. Snoek, C.G.M., Worring, M., van Gemert, J.C., Geusebroek, J.M., Smeulders, A.W.M.: The challenge problem for automated detection of 101 semantic concepts in multimedia. In: Proceedings of 14th ACM International Conference on Multimedia, MULTIMEDIA'06, pp. 421–430 (2006)
49. Sobol-Shikler, T., Robinson, P.: Classification of complex information: Inference of co-occurring affective states from their expressions in speech. IEEE Trans. Pattern Anal. Mach. Intell. **32**(7), 1284–1297 (2010)
50. Sriram, B., Fuhry, D., Demir, E., Ferhatosmanoglu, H., Demirbas, M.: Short text classification in twitter to improve information filtering. In: Proceedings of 33rd international ACM SIGIR conference on Research and development in information retrieval, pp. 841–842. ACM (2010)
51. Sun, L., Ji, S., Ye, J.: Multi-label dimensionality reduction. CRC Press (2013)
52. Tenenboim-Chekina, L., Rokach, L., Shapira, B.: Identification of label dependences for multi-label classification. In: Proceedings of 2nd International Workshop on Learning from Multi-Label Data, MLD'10, pp. 53–60 (2010)
53. Tsoumakas, G., Katakis, I., Vlahavas, I.: Mining multi-label data. In: Data Mining and Knowledge Discovery Handbook, pp. 667–685. Springer (2010)
54. Tsoumakas, G., Xioufis, E.S., Vilcek, J., Vlahavas, I.: MULAN multi-label dataset repository. http://mulan.sourceforge.net/datasets.html
55. Tsoumakas, G., Xioufis, E.S., Vilcek, J., Vlahavas, I.: MULAN: a java library for multi-label learning. J. Mach. Learn. Res. **12**, 2411–2414 (2011)
56. Turnbull, D., Barrington, L., Torres, D., Lanckriet, G.: Semantic annotation and retrieval of music and sound effects. IEEE Trans. Audio Speech Lang. Process. **16**(2), 467–476 (2008)
57. Vembu, S., Gärtner, T.: Label ranking algorithms: a survey. In: Preference learning, pp. 45–64. Springer (2011)
58. Wieczorkowska, A., Synak, P., Raś, Z.: Multi-label classification of emotions in music. In: Intelligent Information Processing and Web Mining. AISC, vol. 35, Chap. 30, pp. 307–315 (2006)
59. Wolpert, D.H., Macready, W.G.: No free lunch theorems for optimization. IEEE Trans. Evol. Comput. **1**(1), 67–82 (1997)
60. Zhang, M., Zhang, K.: Multi-label learning by exploiting label dependency. In: Proceedings of 16th International Conference on Knowledge Discovery and Data Mining, ACM SIGKDD'10, pp. 999–1008 (2010)

Chapter 3
Case Studies and Metrics

Abstract Multilabel classification techniques have been applied in many real-world situations in the last two decades. Each one represents a different case study for MLC, using one or more MLDs. After the general overview provided in Sect. 3.1, this chapter begins by briefly describing in Sect. 3.2 the most usual case studies found in the literature. As a result, a full list of available MLDs will be obtained, and the usual characterization metrics are explained and put in use with them in Sect. 3.3. Then, a practical use case is detailed in Sect. 3.4, running a simple MLC algorithm over a few MLDs. Lastly, the usual performance evaluation metrics for MLC are introduced in Sect. 3.5 and they are used to analyze the results obtained from this experiment.

3.1 Overview

The main application fields of MLC were introduced in the previous chapter from a global perspective. The goal in this chapter was to delve into each one of these fields, enumerating every one of the publicly available MLDs and stating where they come from. In addition to this basic reference information, it would be interesting to get some general characteristics for each MLD. For doing so, most of the characterization metrics described in the literature are going to be introduced, along with their formulations and discussion about their usefulness. Several extensive tables containing each measurement for every MLD will be provided.

In the following chapters, several dozens of MLC algorithms will be described, and some of them will be experimentally tested. Therefore, how to conduct such an experiment, and the way the results can be assessed to evaluate the algorithms' performance, are fundamental aspects. Once the available MLDs and their main traits are known, a basic kNN-based MLC algorithm is introduced and it is run to process some of these MLDs.

Multilabel predictive performance evaluation metrics have to deal with the presence of multiple outputs, taking into consideration the existence of predictions which are partially correct or wrong. As will be expounded, these metrics can be grouped

© Springer International Publishing Switzerland 2016
F. Herrera et al., *Multilabel Classification*,
DOI 10.1007/978-3-319-41111-8_3

into several categories according to distinct criteria. Then, most of the MLC evaluation metrics are explained along with their formulation, using them to assess the
results obtained from the previous experiments.

3.2 Case Studies

In the previous chapter, the main application fields for MLC were portrayed. Attending to the grouping criterion then established, in this section most of the case studies
found in the specialized literature will be enumerated. Table 3.1 summarizes these
case studies, giving their original references and the place they can be downloaded
from.[1]

Some of these case studies have associated several MLDs, whose names and
characteristics will be analyzed later. The same MLD can be available in different
formats,[2] for instance MULAN, MEKA, and KEEL, depending on the repository the
user refers to.

The following subsections cover each MLC application field. The case studies
are alphabetically enumerated inside each category conforming to the name of the
MLD or set of MLDs belonging to them.

3.2.1 Text Categorization

Categorizing text documents into one or more categories is a very usual need. It is
the task at the root of MLC. This is the reason for the existence of many datasets
associated with this use case. The case studies mentioned below have been used in a
considerable portion of the multilabel literature. Some of them have associated more
than one MLD.

- 20ng: This dataset, known as *20 Newsgroups*, has its origin in the task [28] of
  filtering news group messages. The dataset has become a classical problem for
  testing text clustering and text-labeling algorithms. It contains over a thousand
  entries for each one of 20 different news groups, making a total of almost 20 000
  data instances. Some of the news groups are closely related, so some messages
  were cross-posted to more than one group. The input attributes, there are more
  than a thousand, are the terms extracted from all the documents. For each instance,
  those terms appearing in the message are set to 1, while the others are set to 0.
  This representation is known as boolean bag-of-words (BoW) model. There are

---

[1] All datasets are available at RUMDR (*R Ultimate Multilabel Dataset Repository*) [10], from which
can be downloaded and exported to several file formats.

[2] The differences among the main file formats, all of them derived from the ARFF format used by
WEKA, and how to use each of them, will be detailed in Chap. 9.

Table 3.1 Case studies and their categories and references

Case study	Category	References	Download
20ng	Text	[28]	[33]
bibtex	Text	[26]	[3, 43]
birds	Sound	[7]	[43]
bookmarks	Text	[26]	[3, 43]
cal500	Sound	[44]	[43]
corel	Image	[5, 20]	[3, 43]
delicious	Text	[40]	[3, 43]
emotions	Sound	[48]	[3, 33, 43]
enron	Text	[27]	[3, 33, 43]
EUR-Lex	Text	[30]	[43]
flags	Image	[24]	[43]
genbase	Gen/Bio	[19]	[3, 43]
imdb	Text	[32]	[33]
langlog	Text	[31]	[33]
mediamill	Video	[35]	[3, 9, 43]
medical	Text	[18]	[3, 33, 43]
nus-wide	Image	[17]	[43]
ohsumed	Text	[25]	[33]
rcv1v2	Text	[29]	[3, 9, 43]
reuters	Text	[31]	[33]
scene	Image	[6]	[3, 9, 33, 43]
slashdot	Text	[32]	[33]
stackexchange	Text	[15]	[12]
tmc2007	Text	[37]	[9, 43]
yahoo	Text	[47]	[43]
yeast	Gen/Bio	[21]	[3, 9, 33, 43]

20 labels, corresponding to the news groups the messages have been taken from. Only a handful of instances are assigned to more than one label.

- bibtex: Introduced in [26] as part of a tag recommendation task, it contains the metadata for bibliographic entries. The words that presented in the papers' title, authors names, journal name, and publication date were taken as input attributes. The full vocabulary consisted in 1 836 features. The data origin is Bibsonomy,[3] a specialized social network where the users can share bookmarks and BibTeX entries assigning labels to them. bibtex is the dataset generated from the data contained in the BibTeX entries, being associated with a total of 159 different

---

[3]http://www.bibsonomy.org.

labels. The boolean BoW model is used to represent the documents, so all features are binary indicating if a certain term is relevant to the document or not.

- `bookmarks`: This MLD comes from the same source [26] that the previous one. In this case, the data are obtained from the bookmarks shared by the users. Specifically, the URL of the resource, its title, date, and description are included into the dataset. The vocabulary consisted in 2 150 different terms, used as input features. The tags assigned to the bookmarks by the users, a total of 208, are taken as labels. The main difference between `bibtex` and `bookmarks` is the size of the MLD, having the latter more than ten times the number of instances that the former.

- `delicious`: The authors of this dataset [40] are the same of the previous one, and its nature is also similar to `bookmarks`. This time the links to Web pages were taken from the del.icio.us[4] portal. The page content for a set of popular tags was retrieved and parsed, and the resulting vocabulary was filtered to avoid non-frequent words. As a result, an MLD with almost a thousand labels was generated. The goal of the authors was proposing an MLC method able to deal with a so large number of labels.

- `enron`: The Enron corpus is a large set of email messages, with more than half a million entries, from which a dataset for automatic folder assignment research was generated [27]. The `enron` MLD is a subset of the previous dataset, with only 1 701 instances. Each one has as input features a BoW obtained from the email's fields, such as the subject and the body of the message. The labels correspond to the folders in which each message was stored into by the users. A total of 53 labels are considered.

- `EUR-Lex`: This case study is made up of three MLDs. The primary source is the European Union's database of legal documents, which includes laws, agreements, and regulations. Each document is classified in accordance with three criteria, EUROVOC descriptors, directory codes, and subject matters. For doing so, the header of the document indicates which descriptors, codes, and matters are relevant. Therefore, there are three multilabel tasks to accomplish. From this database, the authors of [30] generated the `eurlex-dc`, `eurlex-ev`, and `eurlex-sm` MLDs.[5] Unlike in most cases, reduction techniques were not applied aiming to obtain a limited number of labels. As a result, the `eurlex-ev` MLD has almost 4 000 of them. The three datasets have the same instances with the same set of 5 000 input features. These contain, in the version used in [30], the TF-IDF representation instead of BoW as the previous ones.

- `imdb`: The aim of this study [32] was to automatically classify movies into the proper genres, i.e., drama, comedy, adventure, or musical, among others. A total of 28 genres are considered. The input features were generated from the text gathered from the IMDB[6] database for each movie, relying in a boolean BoW

---

[4]https://delicious.com/.

[5]Additional information about how these MLDs were produced, including the software to do so, can be found at http://www.ke.tu-darmstadt.de/resources/eurlex.

[6]http://imdb.org.

representation. These texts contained a summary of the movies' plot, with a vocabulary made up of a thousand terms. Containing more than 120 000 instances, it is one of the largest MLDs publicly available.

- langlog: Introduced in [31], this MLD has been created from the posts published into the Language Log Forum,[7] a Web site for discussing language-related topics. As many other text MLDs, this also follows the boolean BoW model, with a total of 1 460 input features. The blog entries are categorized by 75 different labels.

- medical: The documents processed to produce this MLD are anonymized clinical texts, specifically the free text where the patient symptoms are described. A portion of the total corpus described in [18] was used to generate the MLD, with the text transformed into a BoW per document. The labels, a total of 45, are the codes from the International Classification of Diseases, precisely ICD-9-CM[8] codes.

- ohsumed: The origin of this dataset [25] is the Medline database, a text corpus from almost three hundred medical journals. The Ohsumed collection is a subset of the Medline dataset compiled in the Oregon Health Science University. The title and abstract texts of each article were processed and represented as BoW, producing a set of thousand input features. Each document is linked to one or more of the 23 main categories of the MeSH diseases ontology.[9] These categories are the labels appearing in the 13 929 instances that the MLD consists in.

- rcv1v2: This case study consists of five MLDs, being each one of them a subset of the original RCV1-v2 (*Reuters Corpus Volume 1 version 2*). The RCV1 text corpus was generated from the full text of English news published by Reuters along one year, from August 20, 1996, to August 19, 1997. Version 2 of this corpus is a corrected version introduced in [29]. Each entry was classified according to three categories, such as topic codes, industry codes, and region codes. A total of 101 different labels are considered. The vocabulary used as input features has 47 236 terms, represented as TF-IDF values. The full RCV1 corpus had 800 000 documents. 6 000 of them are provided in each one of the five subsets.

- reuters: Introduced in [31], it is also a subset of the RCV1 corpus. In this case, a feature selection method has been applied, taking only 500 input attributes instead of the more than 47 000 in rcv1v2. The goal was to work with more representative features. At the same time, the reduced set of attributes improves the speed of the learning process.

- slashdot: The source this MLD was generated from is Slashdot,[10] a well-known news portal mainly focused in technology and in science. The MLD was generated [32] taking the text from the news title and summary, producing a boolean BoW for each entry. The vocabulary has 1 079 terms. The tags used for categorize these entries, a total of 22, act as labels.

- stackexchange: The case study faced in [15] is a tag suggestion task for questions posted in specialized forums, specifically forums from the Stack Exchange

---

[7]http://languagelog.ldc.upenn.edu/nll/.

[8]http://www.cdc.gov/nchs/icd/icd9cm.htm.

[9]https://www.nlm.nih.gov/mesh/indman/chapter_23.html.

[10]http://slashdot.org.

network.[11] Six MLDs were generated from six different forums, devoted to topics such as cooking, computer science, and chess. The title and body of each question was text-mined, producing a frequency BoW. The tags assigned by the users to their questions were used as labels. The vocabulary for each forum is specific, being made of between 540 and 1 763 terms. These worked as input attributes. The labels are specific as well, ranging its number from 123 to 400.

- tmc2007: This dataset bore as a result of the SIAM Text Mining Workshop[12] in 2007 [37]. As many other text datasets, boolean BoW was chosen as a way of representing the terms appearing in documents. Those were aviation safety reports, in which certain problems during flights were described. The vocabulary consists of 49 060 different words, used as input features. Each report is tagged into one or more categories from a set of 22. These are the labels in the MLD.

- yahoo: The authors of [47] compiled for their study the Web pages referenced in 11 out of the 14 main categories of the classical Yahoo![13] Web index. Therefore, 11 MLDs are available for this case study. All of them use the boolean BoW representation, with features obtained from the pages referenced in the index. The number of words goes from 21 000 to 52 000, depending on the MLD. The subcategories that the pages belong to are used as labels. The number of labels is in the range 21–40.

3.2.2 Labeling of Multimedia Resources

Although text resources were the first ones to demand automated classification mechanisms, recently the need for labeling other kind of data, such as images, sounds, music, and video, has experimented a huge growth. By contrast with the case studies enumerated in the previous section, in which a common representation as BoW (whether they contain boolean values, frequencies, or TF-IDF values) is used, the following ones resort to disparate embodiments.

- birds: This MLD emerges from the case study described in [7], where the problem of identifying multiple birds species from acoustic recordings is tackled. The researchers used hundreds of sound snippets, recorded in nature at times of day with high bird activity. Between 1 and 5 different species appear in each snippet. The audio was processed with a 2D time-frequency segmentation approach, aiming to separate syllables overlapping in time. As a result, a set of features with the statistic profile of each segment is produced. Since a sound can be made up of several segments, the produced dataset is a multiinstance multilabel dataset.

- cal500: Tagging music tracks with semantic concepts is the task faced in [44], from which the cal500 MLD is derived. The researchers took five hundred

---

[11] http://stackexchange.com/.

[12] http://web.eecs.utk.edu/events/tmw07/.

[13] http://web.archive.org/web/19970517033654/http://www9.yahoo.com/.

songs, from unique singers, and defined a vocabulary aimed to define aspects such as the emotions produced by the song, the instruments and vocal qualities, and music genre. These concepts, a total of 174, are used as labels. Each music track is assigned at least 3 of them and the average is above 26, which is a quite high number in the multilabel context. The input features were generated by sound segmentation techniques. A distinctiveness of this MLD is that no two instances are assigned the same combination of labels.

- corel: The original Corel dataset was used in two different case studies [5, 20] by the same authors, from which several MLDs have been obtained. The Corel dataset has thousands of images categorized into several groups. In addition, each picture is assigned a set of words describing its content. These pictures were segmented by the authors using the normalized cuts method, generating a set of blobs associated with one or more words. The input features, 500 in total, are the vectors resulting from the segmentation process. In [20] (corel5k), 5 000 instances were taken and there are 374 labels, since a minimum of occurrences was not established. The posterior study in [5] (corel16k) used 138 111 instances grouped into 10 subsets. A minimum of occurrences for each label was set, limiting its number to 153–174 depending on the subset.

- emotions: The origin of this dataset is the study conducted in [48], whose goal is to automatically identify the emotions produced by different songs. A hundred songs from each one of seven music styles were taken as input. The authors used the software tool described in [46] to extract from each record a set of rhythmic features and another one with timbre features. The union of these sets, after a process of feature selection, is used as input attributes. The songs were labeled by three experts, using the six main emotions of the Tellegen-Watson-Clark abstract emotional model. Only those songs where the assigned labels coincide were retained, reducing the number of instances from the original 700 to 593.

- flags: This MLD is considered as a toy dataset, since it only has 194 instances with a set of 19 inputs features and 7 labels. The original version can be found in the UCI repository.[14] In [24] several of its attributes, the ones indicating which colors appear in the flag or if it contains a certain image or text, were defined as labels. The remainder attributes, including the zone and land mass the country belongs to, its area, religion, population, etc., are established as input features.

- mediamill: It was introduced in [35] as a challenge for video indexing. The data consist of a collection of video sequences, taken from the TREC Video Retrieval Evaluation,[15] from which a set of 120 features have been extracted. This set of features is the concatenation of several similarity histograms extracted from the pixels of each frame. The goal was to discover what semantic concepts are associated with each entry, among a set of 101 different labels. Some of these concepts refer to environments, such as road, mountain, sky, or urban, others to physical objects, such as flag, tree, and aircraft. A visual representation of these 101 concepts can be fond in [35].

---

[14] https://archive.ics.uci.edu/ml/datasets/Flags.

[15] http://www-nlpir.nist.gov/projects/trecvid/.

- `nus-wide`: The famous Flickr[16] social network, in which millions of users publish their photographs every day, is the origin for the NUS-WIDE dataset, created by NUS's Lab for Media Search. Each image was segmented extracting color histogram, correlation histogram, edge direction, textures, etc. The resulting MLD has 269 648 instances, and two versions of the MLD with different features representation are available. The first one, known as `nus-wide-BoW`, used clustering to produce a 500 dimensional vector of visual words (real values). The second one, named `nus-wide-VLAD`, the vectors have 128 dimensions and are encoded as cVLAD+ features [36] (real values). In both, each instance has an initial attribute containing the name of the file where the image was stored into. Each image was manually annotated using a 81 items vocabulary, with terms such as animal, house, statue, and garden. These are the labels of the MLD.
- `scene`: This MLD is also related to image labeling, specifically to scene classification. The set of pictures was taken from the Corel dataset and some personal ones by the authors [6] were also included. The MLD is made up of 400 pictures for each main concept, beach, sunset, field, fall foliage, mountain, and urban. Therefore, six non-exclusive labels are considered. The images are transformed to the CIE Luv color space, known for being perceptually uniform, and latter segmented into 49 blocks, computing for each one of them values such as the mean and variance. The result is a vector of 294 real-value features in each instance.

3.2.3 Genetics/Biology

This is the area with less publicly available datasets, which is not surprising due to its complexity. There are two MLDs, one focused in predicting the class of proteins and another one for classifying genes in line with their functional expression.

- `genbase`: The authors of [19] produced this MLD compiling information for 662 different proteins. The Prosite access number[17] was used to identify the 1 185 motif patterns and profiles used as input features. All of them are nominal, taking only the YES or NO values. This way the motifs and profiles present in each protein are indicated. 27 different protein classes are considered, being each protein associated with one or more of them. The PDOC protein class identifiers are used as label names. Something to be taken into account while using this MLD is the presence of one additional feature, the first one, that uniquely identifies each instance.
- `yeast`: In this case [21], the goal was to predict the functional expression for a set of genes. The input features for each gene come from microarray expression data, with a 103 real values vector per instance. A subset of 14 functional classes, whose origin is the Comprehensive Yeast Genome Database,[18] are selected and

---

[16]https://www.flickr.com/.

[17]http://prosite.expasy.org/prosite.html.

[18]http://www.ncbi.nlm.nih.gov/pubmed/15608217.

used as labels. Since each gene can express more than one function at once, in fact this is the usual situation, the result is a dataset with multilabel nature.

3.2.4 Synthetic MLDs

Even though there are a quite large collection of MLDs publicly available, in some situations it can be desirable to work with datasets that have certain characteristics. For instance, if we were designing an algorithm to deal with noisy data it would be interesting to test it with MLDs having different noise levels. This is a trait that could be modeled by generating custom synthetic datasets.

Despite the aforementioned need, which has been demanded by several authors in some papers, there is a lack of tools to produce synthetic MLDs when compared with utilities with the same aim for traditional classification. In most cases, internal programs are used to generate these artificial datasets, and only the characteristics of the data are explained. Fortunately, there are some exceptions, such as the Mldatagen program[19] described in [38].

Since they are created by a program, an a priori limit in the number of MLDs that can be created does not exist. They can hold any number of instances, attributes, and labels but, unlike the enumerated in the previous sections, they do not represent any real situation.

3.3 MLD Characteristics

Before attempting to build a classification model to solve a specific problem, it is important to analyze the main characteristics of the data available to accomplish this task. Understanding the inner traits of the data usually will allow the selection of the best algorithm, parameters, etc. Revealing these traits is the aim of the specific characterization metrics for MLDs defined in the literature.

In the following subsections, many of the available characterization metrics are defined, providing their mathematical expressions, and detailing their usefulness. Many of them will be further applied to the MLDs associated with the previous case studies, and certain facts will be discussed. The nomenclature stated in Sect. 2.2 will be used in all equations.

---

[19]http://sites.labic.icmc.usp.br/mldatagen/.

3.3.1 Basic Metrics

The main difference between traditional and multilabel classification comes from the fact that in the latter each instance is associated with a set of labels. This is the reason behind the first specific metrics designed for MLDs, whose purpose is to assess the *multilabelness* of the data, in other words determining the extent at which the samples in the dataset have more than one label.

An obvious way to calculate such a measure consists in counting the number of labels relevant to every instance in the dataset, then averaging the sum to know the mean number of labels per instance. This simple metric was introduced in [39] as label cardinality or simply *Card* (3.1).

$$Card\,(D) = \frac{1}{n} \sum_{i=1}^{n} |Y_i| \tag{3.1}$$

In this context, n denotes the number of instances in the MLD D, Y_i the labelset of the ith instance, and k the total number of labels considered in D. The higher is the *Card* level, the larger is the number of active labels per instance. As a consequence, MLDs with low *Card* values, near 1.0, would denote that most of its samples have only one relevant label. Therefore, it would be a dataset with little multilabelness nature. On the opposite side, high *Card* values state that the data are truly multilabeled. As a general rule, high *Card* values are linked to MLDs which have large sets of labels, yet the contrary is not always true.

Since *Card* is a metric influenced by the size of the set of labels used by each MLD, and it is expressed using the number of labels as measurement unit, a normalized version (3.2) was also proposed. By dividing *Card* by the number of labels in the MLD, a dimensionless metric, known as label density (*Dens*), is obtained. Usually, a high *Dens* value indicates that the labels in the MLD are well represented in each instance. By contrast, low *Dens* values denote more dispersion, with only a small subset of the labels present in most instances.

$$Dens\,(D) = \frac{1}{k} \frac{1}{n} \sum_{i=1}^{n} |Y_i| \tag{3.2}$$

Another way of assessing the multilabelness of a dataset would be by means of the P_{min} metric (3.3) introduced in [45]. This is simply the percentage of instances in the MLD with only one active label. Intuitively, a high P_{min} value would denote that a large proportion of instances are single labeled.

$$P_{min}\,(D) = \sum_{y' \in Y/|y'|=1} \frac{|y'|}{n} \tag{3.3}$$

Subsets of the labels in the set $\mathcal{L}$ appear in the instances of D forming labelsets. Theoretically 2^k different labelsets could exists, but in practice the number of unique (distinct) labelsets is limited by the number of instances in D. Thus, the number of unique combinations is limited by the expression $min(n, 2^k)$. The effective number of distinct labelsets in an MLD is an indicator of the uniformity in the labels distribution among the samples. The higher the number is, the more irregularly the labels appear in the data instances. The number of distinct labelsets is also known as label diversity (*Div*), and it can also be normalized dividing it by the number of instances.

Furthermore, the frequency of each labelset appears in the MLD may be also an interesting information. Even though the total number of distinct labelsets is not high, if many of them only appear once, associated with one instance, this could lead to some difficulties during the learning process. In addition, the analysis of the labelsets provides information related to dependencies among labels.

Besides the previous ones, in [11] other standard statistical metrics, such as the coefficient of variation, kurtosis, and skewness, are used to characterize how the labels in an MLD are distributed. The joint use of all these metrics can help in gaining insight into this problem.

3.3.2 Imbalance Metrics

The presence of class imbalance in a dataset is a challenge for most learning algorithms. This problem will be analyzed in Chap. 8 in the context of MLC. As will be seen, most MLDs suffer from label imbalance. This means that some labels are much more frequent than others, and it is being an aspect interesting to appraise due to its impact in classification results. Three different metrics to assess label imbalance are proposed in [14], named *IRLlbl* (3.4), *MaxIR* (3.5) and *MeanIR* (3.6). In (3.4), the operator $[\![expr]\!]$ denotes de Iverson bracket. It will return 1 if the expression inside is true or 0 otherwise.

$$IRLbl(l) = \frac{\max\limits_{l' \in L} \left(\sum\limits_{i=1}^{n} [\![l' \in Y_i]\!] \right)}{\sum\limits_{i=1}^{n} [\![l \in Y_i]\!]}. \tag{3.4}$$

With the *IRLbl* metric, it is possible to know the imbalance level of one specific label. This is computed as the proportion between the number of appearances of the most common label and the considered label. Therefore, for the most common label *IRLbl* = *1*. For least frequent labels, the level always will be greater than 1. The higher the *IRLbl*, the rarer is the label presence in the MLD. The goal of the *MaxIR* metric was obtaining the maximum imbalance ratio. In other words, the proportion of the most common label against the most rare one.

$$MaxIR = \max_{l \in L} (IRLbl(l)) \tag{3.5}$$

$$MeanIR = \frac{1}{k} \sum_{l \in L} IRLbl(l). \tag{3.6}$$

Usually a global assessment of the imbalance in the MLD is desired. This metric, named *MeanIR*, is calculated by averaging the *IRLbl* of all labels. Despite the usefulness of this metric by itself, some dispersion measure, such as standard deviation or coefficient of variation, should also be included. A high *MeanIR* could be due to a relatively high *IRLbl* for several labels, but also by cause of extreme imbalance levels for only some labels. In this context, the *CVIR* (3.7) metric provides the additional information needed to know the cause.

$$CVIR = \frac{IRLbl\sigma}{MeanIR}, \quad IRLbl\sigma = \sqrt{\frac{1}{k-1} \sum_{l \in L} (IRLbl\,(l) - MeanIR)^2} \tag{3.7}$$

3.3.3   Other Metrics

Besides the already aforementioned, some other characterization metrics have been proposed in the literature to assess specific qualities of the MLDs. In [13], the *SCUMBLE* metric is introduced as a way to measure the concurrence among very frequent and rare labels. A score is individually computed for each instance (3.8). This score is based on the Atkinson index [4] and the *IRLbl* metric introduced in the previous section. The former is an econometric measure aimed to evaluate income inequalities among the population. In this context, monetary quantities have been replaced by imbalance ratios, provided by the *IRLbl* metric. The result is a value in the [0, 1] range indicating if all the labels in the instance have similar frequencies in the MLD, low values, or by the contrary there are significant differences, the result would be a higher value. The global *SCUMBLE* measure (3.9) is obtained by averaging the score for all instances in the MLD. How these metrics have been the foundation for developing new MLC algorithms will be explained in Chap. 8. As a general rule, higher *SCUMBLE* values denote harder MLDs to learn from.

$$SCUMBLE_{ins}\,(i) = 1 - \frac{1}{IRLbl_i} \left(\prod_{l \in L} IRLbl_{il} \right)^{(1/k)} \tag{3.8}$$

$$SCUMBLE\,(D) = \frac{1}{n} \sum_{i=1}^{n} SCUMBLE_{ins}\,(i) \tag{3.9}$$

The *TCS* (3.10) metric is presented in [16] aiming to facilitate a theoretical complexity indicator. It is calculated as the product of the number of input features, number of labels, and number of different label combinations. To avoid working with very large values, whose interpretation and comparison would be not easy, the *log* function is used to adjust the scale of the previous product. The goal was to determine which MLDs would present a harder work to the preprocessing an learning algorithms. Unlike *SCUMBLE*, *TCS* values are not upper bounded. The higher the value, the more complex would be to process the MLD.

$$TCS(D) = \log(f \times k \times |unique\ labelsets|) \tag{3.10}$$

3.3.4 Summary of Characterization Metrics

Once the main characterization metrics have been defined, they can be used to analyze the MLDs corresponding to the case studies enumerated in Sect. 3.2. Tables 3.2, 3.3, and 3.4 summarize most of these metrics for the MLDs corresponding to case studies from the text, multimedia, and genetics fields, respectively. The columns show, from left to right, **Dataset**: name of the MLD, **n**: number of instances, **f**: number of input attributes, **k**: number of labels, **LSet**: number of distinct labelsets, **Card**: label cardinality (*Card*), **Dens**: label density (*Dens*), **MeanIR**: mean imbalance ratio (*MeanIR*), and **SCUMBLE**: imbalanced labels concurrence level (*SCUMBLE*).

The MLDs from text case studies clearly share a common trait, as almost all of them have a high number of input features, in the range of thousands of them with few exceptions. This is due to the techniques used to mining the text, which produce large collections of words and their frequencies. Many of them also have several hundreds of labels. This, when combined with a large number of instances, also produces a huge amount of labelsets. It is the case with MLDs such as `bookmarks` or `delicious`. Comparatively, the number of features, labels, and labelsets is lower in the datasets coming from multimedia and genetics case studies.

Regarding the *Card* metric that indicates the mean number of labels per instance, most MLDs are in the [1, 5] interval. Some MLDs, such as `20ng`, `langlog`, `slashdot`, `yahoo-reference`, `birds`, and `scene`, are only slightly above 1, meaning that most of its instances are associated with only one label. These would be the less representative cases of what should be a multilabel scenario, since they are closer to a multiclass one. There are a pair of extreme cases in the opposite side. The *Card* values for `delicious` and `cal500` are above 19 and 26, respectively. These MLDs are truly multilabel, with a remarkable average number of active labels in each instance. Halfway between the previous utmost scenarios, the remainder MLDs present the most typical *Card* values, between 2 and 5 labels per instance in average.

Dens is a metric closely related to *Card*. In general, most MLDs have *Dens* values below 0.1. Only those with a very limited set of labels, such as `emotions`, `flags`, or `scene`, or a very high *Card*, such as `cal500`, get a high label density. Therefore,

Table 3.2 Main characteristics of MLDs from text classification case studies

Dataset	n	f	k	LSet	Card	Dens	MeanIR	SCUMBLE
20ng	19 300	1 006	20	55	1.029	0.051	1.007	0.000
bibtex	7 395	1 836	159	2 856	2.402	0.015	12.498	0.094
bookmarks	87 856	2 150	208	18 716	2.028	0.010	12.308	0.060
delicious	16 105	500	983	15 806	19.017	0.019	71.052	0.532
enron	1 702	1 001	53	753	3.378	0.064	73.953	0.303
eurlex-dc	19 348	5 000	412	1 615	1.292	0.003	268.930	0.048
eurlex-ev	19 348	5 000	3 993	16 467	5.310	0.001	396.636	0.420
eurlex-sm	19 348	5 000	201	2 504	2.213	0.011	536.976	0.182
imdb	120 919	1 001	28	4 503	2.000	0.071	25.124	0.108
langlog	1 460	1 004	75	304	1.180	0.016	39.267	0.051
medical	978	1 449	45	94	1.245	0.028	89.501	0.047
ohsumed	13 929	1 002	23	1 147	1.663	0.072	7.869	0.069
rcv1subset1	6 000	47 236	101	1 028	2.880	0.029	54.492	0.224
rcv1subset2	6 000	47 236	101	954	2.634	0.026	45.514	0.209
rcv1subset3	6 000	47 236	101	939	2.614	0.026	68.333	0.208
rcv1subset4	6 000	47 229	101	816	2.484	0.025	89.371	0.216
rcv1subset5	6 000	47 235	101	946	2.642	0.026	69.682	0.238
reuters	6 000	500	103	811	1.462	0.014	51.980	0.052
slashdot	3 782	1 079	22	156	1.181	0.054	17.693	0.013
stackex-chemistry	6 961	540	175	3 032	2.109	0.012	56.878	0.187
stackex-chess	1 675	585	227	1 078	2.411	0.011	85.790	0.262
stackex-coffee	225	1 763	123	174	1.987	0.016	27.241	0.169
stackex-cooking	10 491	577	400	6 386	2.225	0.006	37.858	0.193
stackex-cs	9 270	635	274	4 749	2.556	0.009	85.002	0.272
stackex-philosophy	3 971	842	233	2 249	2.272	0.010	68.753	0.233
tmc2007	28 596	49 060	22	1 341	2.158	0.098	15.157	0.175
tmc2007-500	28 596	500	22	1 172	2.220	0.101	17.134	0.193
yahoo-arts	74 840	23 146	26	599	1.654	0.064	94.738	0.059
yahoo-business	11 214	21 924	30	233	1.599	0.053	880.178	0.125
yahoo-computers	12 444	34 096	33	428	1.507	0.046	176.695	0.097
yahoo-education	12 030	27 534	33	511	1.463	0.044	168.114	0.042
yahoo-entertainment	12 730	32 001	21	337	1.414	0.067	64.417	0.039
yahoo-health	9 205	30 605	32	335	1.644	0.051	653.531	0.092
yahoo-recreation	12 828	30 324	22	530	1.429	0.065	12.203	0.030
yahoo-reference	8 027	39 679	33	275	1.174	0.036	461.863	0.049
yahoo-science	6 428	37 187	40	457	1.450	0.036	52.632	0.058
yahoo-social	12 111	52 350	39	361	1.279	0.033	257.704	0.049
yahoo-society	14 512	31 802	27	1 054	1.670	0.062	302.068	0.096

Table 3.3 Main characteristics of MLDs from multimedia resources classification case studies

Dataset	n	f	k	LSet	Card	Dens	MeanIR	SCUMBLE
birds	645	260	19	133	1.014	0.053	5.407	0.033
cal500	502	68	174	502	26.044	0.150	20.578	0.337
corel5k	5 000	499	374	3 175	3.522	0.009	189.568	0.394
corel16k001	13 766	500	153	4 803	2.859	0.019	34.155	0.273
corel16k002	13 761	500	164	4 868	2.882	0.018	37.678	0.288
corel16k003	13 760	500	154	4 812	2.829	0.018	37.058	0.285
corel16k004	13 837	500	162	4 860	2.842	0.018	35.899	0.277
corel16k005	13 847	500	160	5 034	2.858	0.018	34.936	0.285
corel16k006	13 859	500	162	5 009	2.885	0.018	33.398	0.290
corel16k007	13 915	500	174	5 158	2.886	0.017	37.715	0.282
corel16k008	13 864	500	168	4 956	2.883	0.017	36.200	0.289
corel16k009	13 884	500	173	5 175	2.930	0.017	36.446	0.298
corel16k010	13 618	500	144	4 692	2.815	0.020	32.998	0.279
emotions	593	72	6	27	1.868	0.311	1.478	0.011
flags	194	19	7	54	3.392	0.485	2.255	0.061
mediamill	43 907	120	101	6 555	4.376	0.043	256.405	0.355
nus-wide-BoW	269 648	501	81	18 430	1.869	0.023	95.119	0.171
nus-wide-VLAD	269 648	129	81	18 430	1.869	0.023	95.119	0.171
scene	2 407	294	6	15	1.074	0.179	1.254	0.000

Table 3.4 Main characteristics of MLDs from genetics/proteomics classification case studies

Dataset	n	f	k	LSet	Card	Dens	MeanIR	SCUMBLE
genbase	662	1 186	27	32	1.252	0.046	37.315	0.029
yeast	2 417	103	14	198	4.237	0.303	7.197	0.104

this value is useful to know how sparse are the labelsets in the MLD. Higher *Dens* values will denote labelsets with more active labels than the lower ones.

As can be stated by glancing at the column with the *MeanIR* values, most MLDs show noteworthy imbalance levels. The mean proportion between the frequency of labels are higher to 1:100 in many cases, with some drastic occasions such as eurlex-sm, yahoo-health, or yahoo-business, whose *MeanIR* is above 500. There is only a handful of MLDs that could be considered as balanced, including 20ng, emotions, flags and scene. How this remarkably high imbalance levels can influence the learning methods, and how this difficulty has been faced in the literature, will be the main topics in Chap. 8.

The right-most column in these three tables shows the *SCUMBLE* value for each MLD. Attending to what was stated in [13], values well above 0.1 in this metric designate MLDs in which a significant proportion of rare labels jointly appear with very frequent ones, in the same instances. As can be seen, this is the case for many

Table 3.5 MLDs sorted according to their theoretical complexity score

Rank	Dataset	TCS	f	k	LSet
1	flags	8.879	19	7	54
2	emotions	9.364	72	6	27
3	scene	10.183	294	6	15
4	yeast	12.562	103	14	198
5	birds	13.395	260	19	133
6	genbase	13.840	1 186	27	32
7	20ng	13.917	1 006	20	55
8	slashdot	15.125	1 079	22	156
9	cal500	15.597	68	174	502
10	medical	15.629	1 449	45	94
11	tmc2007-500	16.372	500	22	1 172
12	langlog	16.946	1 004	75	304
13	ohsumed	17.090	1 002	23	1 147
14	stackex-coffee	17.446	1 763	123	174
15	enron	17.503	1 001	53	753
16	reuters	17.548	500	103	811
17	mediamill	18.191	120	101	6 555
18	imdb	18.653	1 001	28	4 503
19	stackex-chess	18.779	585	227	1 078
20	yahoo-business	18.848	21 924	30	233
21	nuswide-VLDA	19.076	129	81	18 430
22	yahoo-entertainment	19.238	32 001	21	337
23	stackex-chemistry	19.473	540	175	3 032
24	yahoo-health	19.609	30 605	32	335
25	corel16k010	19.638	500	144	4 692
26	yahoo-recreation	19.684	30 324	22	530
27	yahoo-reference	19.702	39 679	33	275
28	yahoo-arts	19.703	23 146	26	599
29	corel16k001	19.722	500	153	4 803
30	corel16k003	19.730	500	154	4 812
31	corel16k004	19.791	500	162	4 860
32	corel16k002	19.805	500	164	4 868
33	corel16k005	19.814	500	160	5 034
34	corel16k006	19.821	500	162	5 009
35	corel16k008	19.847	500	168	4 956
36	stackex-philosophy	19.905	842	233	2 249
37	corel16k009	19.919	500	173	5 175
38	corel16k007	19.922	500	174	5 158
39	yahoo-education	19.956	27 534	33	511

(continued)

Table 3.5 (continued)

Rank	Dataset	TCS	f	k	LSet
40	yahoo-computers	19.993	34 096	33	428
41	corel5k	20.200	499	374	3 175
42	yahoo-science	20.337	37 187	40	457
43	yahoo-social	20.418	52 350	39	361
44	nuswide-BoW	20.433	501	81	18 430
45	stackex-cs	20.532	635	274	4 749
46	bibtex	20.541	1 836	159	2 856
47	yahoo-society	20.623	31 802	27	1 054
48	tmc2007	21.093	49 060	22	1 341
49	stackex-cooking	21.111	577	400	6 386
50	eurlex-sm	21.646	5 000	201	2 504
51	eurlex-dc	21.925	5 000	412	1 615
52	rcv1subset4	22.082	47 229	101	816
53	rcv1subset3	22.223	47 236	101	939
54	rcv1subset5	22.230	47 235	101	946
55	rcv1subset2	22.239	47 236	101	954
56	rcv1subset1	22.313	47 236	101	1 028
57	delicious	22.773	500	983	15 806
58	bookmarks	22.848	2 150	208	18 716
59	eurlex-ev	26.519	5 000	3 993	16 467

of the MLDs shown in the previous tables. Some of them, such as the MLDs coming from the Corel image database, enron and delicious, stand out with *SCUMBLE* values as high as 0.532. This means that those MLDs would be specially harder for preprocessing and learning algorithms.

A metric which does not appear in the previous tables is *TCS*. Since it provides a score of the theoretical complexity of the MLDs, it is more useful to look at it after sorting the MLDs by the *TCS* column, instead of alphabetically. The result is shown in Table 3.5. Along the mentioned score, the number of features, labels, and labelsets are also presented. From this table, it is easy to deduct that some of the MLDs previously described as toy datasets present the lower theoretical complexity, with *TCS* values around 10. Unsurprisingly, the text MLDs appear as the most complex ones, due to their large sets of features and labels. Remember that *TCS* values are logarithmic, so a difference of only one unit implies one order of magnitude lower or higher.

Obviously, the MLDs could also be ordered by their *Card*, *MeanIR*, *SCUMBLE* or any other metric values, depending on which traits of the data the interest is on. It is easy to do so using the tools described in Chap. 9.

3.4 Multilabel Classification by Example

At this point, the source, nature, and main characteristics of a large set of MLDs have been already introduced. The characterization metrics have been applied over the MLDs, obtaining the measures shown in the previous tables. Before going into the study of the evaluation metrics, whose goal was to assess the predictive performance of a classifier, some predictions would be needed. This way we could get a glimpse of the values returned by these metrics. For this reason, this section is devoted to demonstrate how to conduct an example of multilabel classification job.

Even though the description of MLC algorithms is the main topic of further chapters, in the following subsection a specific algorithm is introduced to be able to complete the task. The outputs provided by this algorithm are then evaluated by means of different multilabel evaluation metrics.

3.4.1 The ML-kNN Algorithm

One of the simplest approaches to classification is that of kNN. Once a new data sample is given, a kNN classifier looks for its k-nearest neighbors. For doing so, the distance (in some f-dimensional space) between the features of the new sample and all instances in the dataset is computed. Once the closer instances have been gathered, their classes are used to predict the one for the new sample. Since kNN does not create any model, only when a new sample arrives the classifier does some work, it is usually known as a lazy [1] method. It is also frequently referred as instance-based learning [2].

ML-kNN [49] is an adaptation of the kNN method to the multilabel scenario. Unlike the classic kNN algorithm, ML-kNN is not so lazy. It starts by building a limited model that consists of two pieces of information:

- The a priori probabilities for each label. These are simply the number of times each label appears in the MLD divided by the total number of instances. A smoothing factor is applied to avoid multiplying by zero.
- The conditional probabilities for each label, computed as the proportion of instances with the considered label whose k-nearest neighbors, also have the same label.

These probabilities are independently computed for each label, facing the task as a collection of individual binary problems. Therefore, the potential dependencies among labels are fully dismissed by this algorithm.

After this limited training process, the classifier is able to predict the labels for new instances. When a new sample arrives, it goes through the following steps:

- First, the k-nearest neighbors of the given sample are obtained. By default the $L^2 - norm$ (Euclidean distance) is used to measure the similarity between the reference instance and the samples in the MLD.
- Then, the presence of each label in the neighbors is used as evidence to compute maximum a posteriori (MAP) probabilities from the conditional ones obtained before.
- Lastly, the labelset of the new sample is generated from the MAP probabilities. The probability itself is provided as a confidence level for each label, thus making possible to also generate a label ranking.

The reference MATLAB implementation for the ML-kNN algorithm is supplied by the author at his own Web site.[20] There is also available a Java implementation in MULAN. The latter has been used in order to conduct the experimentation described below.

3.4.2 Experimental Configuration and Results

Five MLDs have been chosen to run the ML-kNN algorithm. Two of them are from the text domain (enron and stackex-cs), two more from the multimedia field (emotions and scene), and the last one comes from the biology domain (genbase). Attending to their *TCS* measure emotions and scene, ranked at positions 2 and 3 in Table 3.5, would be the easier cases. A little harder would be genbase (6th), followed by enron (15th) and finally stackex-cs (45th) which, theoretically, would be the most difficult MLD in this collection.

The MLDs were partitioned following a 2×5 strategy. This means that there are two repetitions with 5 folds, and that for each run 80% (4/5) of instances are used for training and 20% (1/5) for testing. Therefore, a total of 10 runs are made for each MLD. Random sampling was used to select the instances in each fold. The full set of folds for the aforementioned five MLDs is available in the book repository [12].

From each run, a set of predictions are obtained from the classifier. These can be assessed using many performance evaluation metrics (they will be described in the next section), getting a set of values for each metric/fold. These values are then averaged, obtaining the mean indicators which are usually reported in most papers, sometimes along with their deviations. Table 3.6 shows all these values, whose interpretation will be further provided as the evaluation metrics are described.

---

[20]http://cse.seu.edu.cn/people/zhangml/Resources.htm#codes.

Table 3.6 Classification results produced by ML-kNN assessed with several evaluation metrics

	stackex-cs	emotions	enron	genbase	scene
Accuracy ↑	0.0540	0.5391	0.3156	0.9440	0.6667
AvgPrecision ↑	0.3009	0.7990	0.6280	0.9860	0.8648
Coverage ↓	77.9260	1.7715	13.2092	0.6110	0.4797
F-measure ↑	0.5900	0.7776	0.5898	0.9776	0.9593
HammingLoss ↓	0.0091	0.1940	0.0524	0.0048	0.0869
MacroF-measure ↑	0.1999	0.6225	0.4284	0.9357	0.7378
MacroPrecision ↑	0.5866	0.7279	0.5568	0.9795	0.8149
MacroRecall ↑	0.0169	0.5981	0.0808	0.6787	0.6808
MicroAUC ↑	0.8481	0.8565	0.9002	0.9893	0.9405
MicroF-measure ↑	0.1065	0.6652	0.4715	0.9458	0.7331
MicroPrecision ↑	0.6289	0.7217	0.6613	0.9934	0.8137
MicroRecall ↑	0.0583	0.6186	0.3671	0.9031	0.6673
OneError ↓	0.6571	0.2799	0.3070	0.0129	0.2269
Precision ↑	0.6157	0.7182	0.6616	0.9956	0.8252
RLoss ↓	0.1522	0.1608	0.0929	0.0072	0.0786
Recall ↑	0.0582	0.6184	0.3654	0.9454	0.6836
SubsetAccuracy ↑	0.0165	0.2968	0.0564	0.9132	0.6243

The arrow at the right of each metric name indicates whether lower values are better
(↓) or the opposite (↑).

Disparate plot designs can be used to graphically represent those final values,
being bar plots and line plots among the most frequent ones. When the interest is in
comparing a group of cases, in this occasion how the classifier has performed with
each MLD in accordance with several metrics, a radar chart (also known as spider
plot) can be useful. In Figs. 3.1 and 3.2, this type of representation has been used to
show the results produced by ML-kNN. Each vertex corresponds to a metric.[21] The
points belonging to an MLD are connected so that a polygon is generated. The larger
is the area of the polygon, the better is the result with a certain MLD.

Through the observation of these two plots, despite the details of each metric are
not yet known, the following facts can be deducted:

- The performance with `emotions` and `scene`, with share a very similar *TCS*
value, is very much alike.
- The results for he previous two MLDs are clearly better than for `enron`, which
has a higher *TCS* score.
- The worst results are in general attributable to `stackex-cs`, the most complex
MLD according to the *TCS* metric.

---

[21]The values of metrics such as *HammingLoss*, *OneError*, and *RankingLoss* have been complemented as the difference with respect to 1, aiming to preserve the principle of assigning a larger
area to better values.

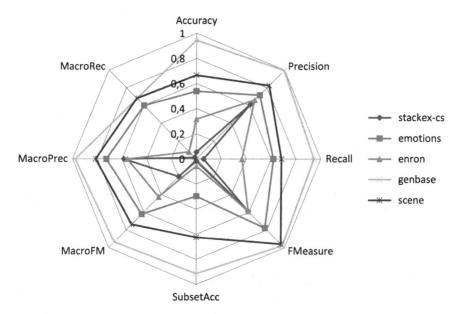

Fig. 3.1 Classification results produced by ML-kNN (part 1)

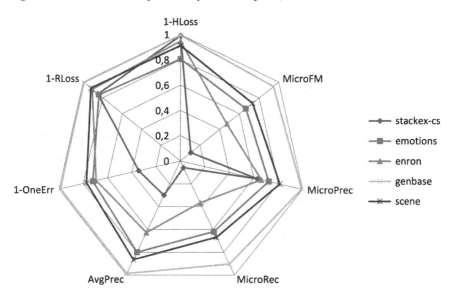

Fig. 3.2 Classification results produced by ML-kNN (part 2)

- The `genbase` results are not in line with previous appraisals, since it obtains the best results in all measures. This could be due to the existence of an attribute, named `protein`, containing a code that uniquely identifies each protein in the MLD. This feature would allow the classifier to easily locate the closest instances, producing a prediction that would be not so precise without that knowledge.

Overall, the previous plots seem to denote that the easier is the MLD, the better the classifier performs. This rule, as can be seen with the exception of `genbase`, can be broken depending on the MLDs specificities. Moreover, these results correspond to one classifier only, so they must be taken cautiously.

In order to complete the judgment of these results, it would be essential to gain an understanding of each individual evaluation metric. In the last section of this chapter, the details regarding how the performance of classifier can be assessed are provided, including additional discussion related to the values in Table 3.6.

3.5 Assessing Classifiers Performance

The output of any multilabel classifier consists of the labelset predicted for each test instance. When working in the traditional scenario, with only one class as output, the prediction only can be correct or wrong. A multilabel prediction, by contrast, can be fully correct, partially correct/wrong (at different degrees), or totally wrong. Applying the same metrics used in traditional classification is possible, but usually it is excessively strict. This is the reason for using specific evaluation metrics, able to take into consideration the cases between the two extremes.

Currently, more than twenty distinct performance metrics have been defined in the literature, and some of them quite specific aimed to hierarchical multilabel classification. All multilabel evaluation metrics can be grouped conforming to two criteria:

- **How the prediction is computed**: A measurement can be made by instance or by label, giving as a result two different groups of metrics:

  - **Example-based metrics**: These metrics [22, 23, 34] are calculated separately for each instance and then averaged dividing between the number of samples.
  - **Label-based metrics**: In contrast to the previous group, the label-based metrics [42] are computed independently for each label before they are averaged. For doing so, two different strategies [41] can be applied:
    Macro-averaging: The metric is calculated individually for each label and the result is averaged dividing by the number of labels (k).
    Micro-averaging: The counters of hits and misses for each label are firstly aggregated, and then the metric is computed only once.

- **How the result is provided**: The output produced by a multilabel classifier can be a binary bipartition of labels or a label ranking. Some of them provide both results.

- **Binary bipartition**: A binary bipartition is a vector of 0s and 1s indicating which of the labels belonging to the MLD are relevant to the processed sample. There are metrics that operate over these bipartitions, using the counters of true positives, true negatives, false positives, and false negatives.
- **Label ranking**: The output is a list of labels ranked according to some relevance measure. A binary bipartition can be obtained from a label ranking by applying a threshold, usually given by the classifier itself. However, there are performance metrics that work with raw rankings to compute the measurement, instead of using counters of right and wrong predictions.

In the two following subsections, the example-based and label-based metrics commonly used in the literature are described, providing their mathematical formulation. Where applicable, each metric description is completed with a discussion of the results produced by the experimentation with ML-kNN in the previous section.

3.5.1 Example-Based Metrics

These are the performance metrics which are firstly evaluated by each instance and then averaged according to the number of instances considered. Therefore, the same weight is assigned to every instance in the final score, whether they contain frequent or rare labels.

3.5.1.1 Hamming Loss

Hamming loss is probably the most commonly used performance metric in MLC. This is not surprising, as it is easy to calculate as can be seen in (3.11). The Δ operator returns the symmetric difference between Y_i, the real labelset of the ith instance, and Z_i, the predicted one. The $|r|$ operator counts the number of 1s in this difference, in other words the number of miss predictions. The total number of mistakes in the n instances is aggregated and then normalized taking into account the number of labels and number of instances.

$$HammingLoss = \frac{1}{n}\frac{1}{k}\sum_{i=1}^{n}|Y_i \Delta Z_i| \qquad (3.11)$$

Since the mistakes counter is divided by the number of labels, this metric will result in different assessments for the same amount of errors when used with MLDs having disparate labelset lengths. This is the main reason for the low *HammingLoss* value of the stackex-cs when compared to emotions or scene. The former has a large number of labels, while the others only have six. Therefore, this metric is an indicator of committed errors by the classifier proportional to the labelset length. We can compare the results of emotions and scene, both have the same number of labels, and conclude that ML-kNN has performed better with the latter (lower value) than the former.

3.5.1.2 Accuracy

In the multilabel field, *Accuracy* is defined as (3.12) the proportion between the number of correctly predicted labels and the total number of active labels, in the both real labelset and the predicted one. The measure is computed by each instance and then averaged, as all example-based metrics.

$$Accuracy = \frac{1}{n} \sum_{i=1}^{n} \frac{|Y_i \cap Z_i|}{|Y_i \cup Z_i|} \tag{3.12}$$

The *Accuracy* for genbase is very high, due to the reason previously explained. As shown in Fig. 3.1, the values for emotions and scene are very similar again, although with a slight advantage to the latter. The obtained *Accuracy* cannot be considered as good in the case of enron, and even less with the stackex-cs MLD. It must be remembered that this MLD had the highest *TCS* of the five case studies. Therefore, that it gets the worst classification performance was within the expected.

3.5.1.3 Precision, Recall, and F-Measure

Precision (3.13) is considered one of the more intuitive metrics to assess multilabel predictive performance. It is calculated as the proportion between the number of labels correctly predicted and the total number of predicted labels. Thus, it can be interpreted as the percentage of predicted labels which are truly relevant for the instance. This metric is usually used in conjunction with *Recall* (3.14) that returns the percentage of labels correctly predicted among all truly relevant labels. That is, the ratio of true labels is given as output by the classifier.

$$Precision = \frac{1}{n} \sum_{i=1}^{n} \frac{|Y_i \cap Z_i|}{|Z_i|} \tag{3.13}$$

$$Recall = \frac{1}{n} \sum_{i=1}^{n} \frac{|Y_i \cap Z_i|}{|Y_i|} \tag{3.14}$$

The jointly use of *Precision* and *Recall* is so common in the information retrieval (IR) field that a metric combining them is defined. It is known as *F-measure* (3.15) and computed as the harmonic mean of the previous ones. This way a weighted measure of how many relevant labels are predicted and how many of the predicted labels are relevant is obtained.

$$F\text{-}measure = 2 * \frac{Precision * Recall}{Precision + Recall}. \tag{3.15}$$

By observing the right side of Fig. 3.1, where *Precision*, *Recall*, and *F-measure* are depicted, that `scene` and `emotions` are once again very close can be stated, though `scene` results are a bit better. With `enron`, it can be seen that *Precision* has a higher value than *Recall*, a far greater fact in the case of `stackex-cs`. This means that for these MLDs a high proportion of the labels included in the prediction are relevant labels, but that there are many other true labels which are not predicted by the classifier. Looking at the *F-measure* values, the same correlation between the theoretical complexity (*TCS* value) of each MLD and classification performance assessment can be deduced.

3.5.1.4 Subset Accuracy

This is possibly the most strict evaluation metric. It is also known as classification accuracy and labelset accuracy, since full labelsets, the predicted and the real one, are compared for full equality as can be seen in (3.16). The larger is the labelset, the lower the likelihood that the classifier produces exactly the correct output. Therefore, for MLDs with large sets of labels that low *SubsectAccuracy* values are obtained is something usual.

$$SubsetAccuracy = \frac{1}{n} \sum_{i=1}^{n} [\![Y_i = Z_i]\!] \tag{3.16}$$

Apart from the atypical case of `genbase`, the *SubsectAccuracy* for the MLDs used in the previous experimentation reflects the problems the classifier had with each one of them. While `scene` values are not bad, the performance with `emotions` was far worse. As could be expected, due to their large sets of labels, `enron` and `stackex-cs` show the worst results.

3.5.1.5 Ranking-Based Metrics

All the example-based metrics described above work over binary partitions of labels, so they need a labelset as output from the classifier. By contrast, the explained here need a ranking of labels, so a confidence degree or belonging probability of each label is needed.

In the following equations, $rank(x_i, l)$ is defined as a function that for the x_i instance and the relevant label $l \in \mathcal{Y}$, whose position is known, returns l's confidence degree into the Z_i prediction returned by the classifier.

The *AvgPrecision* (Average precision) metric (3.17) determines for each label in an instance, the proportion of relevant labels that are ranked above it in the predicted ranking. The goal was to know how many positions have to be checked, in average, before a non-relevant label is found. Therefore, the larger is the *AvgPrecision* measure obtained, the better would be performing the classifier.

$$AveragePrecision = \frac{1}{n} \sum_{i=1}^{n} \frac{1}{|Y_i|} \sum_{y \in Y_i} \frac{|\{y' | rank(x_i, y') \le rank(x_i, y), y' \in Y_i\}|}{rank(x_i, y)}$$

$$(3.17)$$

The *Coverage* metric (3.18) counts the number of steps to going through the ranking provided by the classifier until all the relevant labels are found. The lower is the mean number of steps for the MLD, value returned by *Coverage*, the better is performing the classifier. As can be shown in (3.18), this measure is not normalized, so it is not upper bounded. As happens with other multilabel classification metrics, *Coverage* is influenced for the size of the set of labels in each MLD. The larger is this set, the higher usually is the mean number of steps to walk-through the ranking.

$$Coverage = \frac{1}{n} \sum_{i=1}^{n} \underset{y \in Y_i}{\operatorname{argmax}} \langle rank(x_i, y) \rangle - 1 \qquad (3.18)$$

As the previous one, *OneError* (3.19) is a performance metric to minimize. The expression which follows the summation returns 1 if the top-ranked label in the prediction given by the classifier does not belong to the real labelset. The number of miss predictions is accumulated and averaged. The result is the percentage of cases in which the most confident label for the classifier is a false positive.

$$OneError = \frac{1}{n} \sum_{i=1}^{n} [\![[\underset{y \in Z_i}{\operatorname{argmax}} \langle rank(x_i, y) \rangle \notin Y_i]]\!]. \qquad (3.19)$$

The *RLoss* (Ranking loss) metric takes all possible combinations of relevant and non-relevant labels for an instance and counts (3.20) how many times a non-relevant label is ranked above a relevant one in the classifier prediction. The counting is normalized dividing by the product of relevant and non-relevant labels in the instance and then averaged by the number of assessed instances. The lower is the *RLoss* measure, the better is performing the classifier.

$$RLoss = \frac{1}{n} \sum_{i=1}^{n} \frac{1}{|Y_i|.|\overline{Y_i}|} |y_a, y_b : rank(x_i, y_a) > rank(x_i, y_b), (y_a, y_b) \in Y_i \times \overline{Y_i}|$$

$$(3.20)$$

Observing the *AvgPrecision* in Table 3.6, it can be seen that with the exception of `stackex-cs`, ML-kNN performed quite well with the other four MLDs. Looking at the *Coverage* row, the values for `stackex-cs` and `enron` stand out. Since they have more labels, the number of steps to complete before getting all relevant labels is higher. The *OneError* values are quite similar for `emotions`, `scene` and `enron`, while for `stackex-cs` is much higher. This denotes that for the latter MLD the top-ranked label was not usually relevant. Lastly, considering the *RLoss* values a different scenario is observed. In this case, the worst results are obtained from `emotions`, though `stackex-cs` is very close. Although `emotions` only has six labels, there are a significant amount of predictions made by ML-kNN in which non-relevant labels are ranked above the relevant ones.[22]

3.5.2 Label-based Metrics

All the performance metrics enumerated in the previous section are evaluated individually for each instance, and then averaged dividing by the number of considered instances. Therefore, each data sample is given the same weight in the final result. On the contrary, label-based metrics can be computed by means of two different averaging strategies. These are usually known as *macro-averaging* and *microaveraging*.

Any of the metrics obtained from a binary partition of labels, such as *Precision*, *Recall* or *F-measure*, can be also computed using these strategies. For doing so, the generic formulas in (3.21) and (3.22) are used. *EvalMet* would be one of the metrics just mentioned. In this context, TP stands for *True Positives*, FP for *False Positives*, TN for *True Negatives*, and FN for *False Negatives*.

$$MacroMet = \frac{1}{k} \sum_{l \in \mathcal{L}} EvalMet(TP_l, FP_l, TN_l, FN_l) \qquad (3.21)$$

$$MicroMet = EvalMet(\sum_{l \in \mathcal{L}} TP_l, \sum_{l \in \mathcal{L}} FP_l, \sum_{l \in \mathcal{L}} TN_l, \sum_{l \in \mathcal{L}} FN_l) \qquad (3.22)$$

In the macro-averaging approach, the metric is evaluated once per label, using the accumulated counters for it, and then the mean is obtained dividing by the number of labels. This way the same weight is assigned to each label, whether it is very frequent or very rare.

---

[22] It must be taken into account that ML-kNN does not generate a real ranking of labels as prediction, but a binary partition. The ranking is generated from the posterior probabilities calculated for each label. With so few labels in `emotions`, it is possible to have many ties in these probabilities, so the positions in the ranking could be randomly determined in some cases.

On the contrary, the microaveraging strategy first adds the counters for all labels and then computes the metric only once. Since the predictions where rare labels appear are combined with that made for the most frequent ones, the former are usually diluted among the latter. Therefore, the contribution of each label to the final measure is not the same.

In addition to label-based metrics computed from binary partitions, those calculated from labels rankings are also available. The area under the ROC (*Receiver Operating Characteristic*) curve (AUC) can be computed according to the macro- and (3.23) and micro- (3.24) averaging approaches

$$MacroAUC = \frac{1}{k} \sum_{l \in \mathcal{L}} \frac{|\{x', x'' : rank(x', y_l) \geq rank(x'', y_l), (x', x'') \in X_l \times \overline{X_l}\}|}{|X_l| \cdot |\overline{X_l}|},$$

$$X_l = \{x_i | y_l \in Y_i\}, \overline{X_l} = \{x_i | y_l \notin Y_i\}$$
$$(3.23)$$

$$MicroAUC = \frac{|\{x', x'', y', y'' : rank(x', y') \geq rank(x'', y''), (x', y') \in S^+, (x'', y'') \in S^-\}|}{|S^+| \cdot |S^-|},$$

$$S^+ = \{(x_i, y) | y \in Y_i\}, S^- = \{(x_i, y) | y \notin Y_i\}$$
$$(3.24)$$

Analyzing the results in Table 3.6 corresponding to the label-based metrics, some interesting conclusions can be drawn. The *MacroF-measure* for genbase is clearly under the *MicroF-measure*. In the both cases, the same basic metric is used, *F-measure*, but with a different averaging strategy. From this observation, it can be deducted that one or more miss predicted rare labels exist in this MLD. By looking at Table 3.4 that genbase has a remarkable imbalance level can be confirmed, the existence of some rare labels is a fact. On the other hand, the *MicroAUC* values for all the MLDs are above the 0.8 level, which is the threshold from which usually the results are considered as good. The values for enron, genbase, and scene even surpass the 0.9 limit and can be regarded as excellent.

In addition to the groups of metrics already explained here, several more can be found, in general much more specific, in the specialized literature. For instance, there are metrics for evaluating the performance in hierarchical multilabel classification such as *Hierarchical loss* [8]. It is based on *Hamming loss*, but considering the level of the hierarchy where the miss predictions are made.

References

1. Aha, D.W. (ed.): Lazy Learning. Springer (1997)
2. Aha, D.W., Kibler, D., Albert, M.K.: Instance-based learning algorithms. Mach. Learn. **6**(1), 37–66 (1991)
3. Alcala-Fdez, J., Fernández, A., Luengo, J., Derrac, J., García, S., Sánchez, L., Herrera, F.: KEEL multi-label dataset repository. http://sci2s.ugr.es/keel/multilabel.php
4. Atkinson, A.B.: On the measurement of inequality. J. Econ. Theory **2**(3), 244–263 (1970)
5. Barnard, K., Duygulu, P., Forsyth, D., de Freitas, N., Blei, D.M., Jordan, M.I.: Matching words and pictures. J. Mach. Learn. Res. **3**, 1107–1135 (2003)
6. Boutell, M., Luo, J., Shen, X., Brown, C.: Learning multi-label scene classification. Pattern Recogn. **37**(9), 1757–1771 (2004)
7. Briggs, F., Lakshminarayanan, B., Neal, L., Fern, X.Z., Raich, R., Hadley, S.J.K., Hadley, A.S., Betts, M.G.: Acoustic classification of multiple simultaneous bird species: a multi-instance multi-label approach. J. Acoust. Soc. Am. **131**(6), 4640–4650 (2012)
8. Cesa-Bianchi, N., Gentile, C., Zaniboni, L.: Incremental algorithms for hierarchical classification. J. Mach. Learn. Res. **7**, 31–54 (2006)
9. Chang, C.C., Lin, C.J.: LIBSVM data: multi-label classification repository. http://www.csie.ntu.edu.tw/~cjlin/libsvmtools/datasets/multilabel.html
10. Charte, F., Charte, D., Rivera, A.J., del Jesus, M.J., Herrera, F.: R Ultimate multilabel dataset repository. In: Proceedings of 11th International Conference on Hybrid Artificial Intelligent Systems, HAIS'16, vol. 9648, pp. 487–499. Springer (2016)
11. Charte, F., Rivera, A., del Jesus, M.J., Herrera, F.: LI-MLC: a label inference methodology for addressing high dimensionality in the label space for multilabel classification. IEEE Trans. Neural Netw. Learn. Syst. **25**(10), 1842–1854 (2014)
12. Charte, F., Rivera, A.J., del Jesus, M.J., Herrera, F.: Multilabel classification. Problem analysis, metrics and techniques book repository. https://github.com/fcharte/SM-MLC
13. Charte, F., Rivera, A.J., del Jesus, M.J., Herrera, F.: Concurrence among Imbalanced labels and its influence on multilabel resampling algorithms. In: Proceedings of 9th International Conference on Hybrid Artificial Intelligent Systems, HAIS'14, vol. 8480. Springer (2014)
14. Charte, F., Rivera, A.J., del Jesus, M.J., Herrera, F.: Addressing imbalance in multilabel classification: measures and random resampling algorithms. Neurocomputing **163**, 3–16 (2015)
15. Charte, F., Rivera, A.J., del Jesus, M.J., Herrera, F.: QUINTA: a question tagging assistant to improve the answering ratio in electronic forums. In: Proceedings of IEEE International Conference on Computer as a Tool, EUROCON'15, pp. 1–6. IEEE (2015)
16. Charte, F., Rivera, A.J., del Jesus, M.J., Herrera, F.: On the impact of dataset complexity and sampling strategy in multilabel classifiers performance. In: Proceedings of 11th International Conference on Hybrid Artificial Intelligent Systems, HAIS'16, vol. 9648, pp. 500–511. Springer (2016)
17. Chua, T.S., Tang, J., Hong, R., Li, H., Luo, Z., Zheng, Y.: NUS-WIDE: a real-world web image database from National University of Singapore. In: Proceedings of 8th ACM international Conference on Image and Video Retrieval, CIVR'09, pp. 48:1–48:9. ACM (2009)
18. Crammer, K., Dredze, M., Ganchev, K., Talukdar, P.P., Carroll, S.: Automatic code assignment to medical text. In: Proceedings of Workshop on Biological, Translational, and Clinical Language Processing, BioNLP'07, pp. 129–136. Association for Computational Linguistics (2007)
19. Diplaris, S., Tsoumakas, G., Mitkas, P., Vlahavas, I.: Protein classification with multiple algorithms. In: Proceedings of 10th Panhellenic Conference on Informatics, PCI'05, vol. 3746, pp. 448–456. Springer (2005)
20. Duygulu, P., Barnard, K., de Freitas, J., Forsyth, D.: Object recognition as machine translation: learning a Lexicon for a fixed image vocabulary. In: Proceedings of 7th European Conference on Computer Vision, ECCV'02, vol. 2353, pp. 97–112. Springer (2002)
21. Elisseeff, A., Weston, J.: A kernel method for multi-labelled classification. In: Advances in Neural Information Processing Systems, vol. 14, pp. 681–687. MIT Press (2001)

22. Ghamrawi, N., McCallum, A.: Collective multi-label classification. In: Proceedings of 14th ACM International Conference on Information and Knowledge Management, CIKM'05, pp. 195–200. ACM (2005)
23. Godbole, S., Sarawagi, S.: Discriminative methods for multi-labeled classification. Adv. Knowl. Discov. Data Min. **3056**, 22–30 (2004)
24. Gonçalves, E.C., Plastino, A., Freitas, A.A.: A genetic algorithm for optimizing the label ordering in multi-label classifier chains. In: Proceedings of 25th IEEE International Conference on Tools with Artificial Intelligence, ICTAI'13, pp. 469–476. IEEE (2013)
25. Joachims, T.: Text categorization with suport vector machines: learning with many relevant features. In: Proceedings of 10th European Conference on Machine Learning, ECML'98, pp. 137–142. Springer (1998)
26. Katakis, I., Tsoumakas, G., Vlahavas, I.: Multilabel text classification for automated tag suggestion. In: Proceedings of European Conference on Machine Learning and Principles and Practice of Knowledge Discovery in Databases, ECML PKDD'08, pp. 75–83 (2008)
27. Klimt, B., Yang, Y.: The enron corpus: a new dataset for email classification research. In: Proceedings of 15th European Conference on Machine Learning, ECML'04, pp. 217–226. Springer (2004)
28. Lang, K.: Newsweeder: learning to filter netnews. In: Proceedings of 12th International Conference on Machine Learning, ML'95, pp. 331–339 (1995)
29. Lewis, D.D., Yang, Y., Rose, T.G., Li, F.: RCV1: a new benchmark collection for text categorization research. J. Mach. Learn. Res. **5**, 361–397 (2004)
30. Mencia, E.L., Fürnkranz, J.: Efficient pairwise multilabel classification for large-scale problems in the legal domain. In: Proceedings of 11th European Conference on Machine Learning and Knowledge Discovery in Databases, ECML PKDD'08, pp. 50–65. Springer (2008)
31. Read, J.: Scalable multi-label classification. Ph.D. thesis, University of Waikato (2010)
32. Read, J., Pfahringer, B., Holmes, G., Frank, E.: Classifier chains for multi-label classification. Mach. Learn. **85**, 333–359 (2011)
33. Read, J., Reutemann, P.: MEKA multi-label dataset repository. http://sourceforge.net/projects/meka/files/Datasets/
34. Schapire, R.E., Singer, Y.: Boostexter: a boosting-based system for text categorization. Mach. Learn. **39**(2–3), 135–168 (2000)
35. Snoek, C.G.M., Worring, M., van Gemert, J.C., Geusebroek, J.M., Smeulders, A.W.M.: The challenge problem for automated detection of 101 semantic concepts in multimedia. In: Proceedings of 14th ACM International Conference on Multimedia, MULTIMEDIA'06, pp. 421–430 (2006)
36. Spyromitros-Xioufis, E., Papadopoulos, S., Kompatsiaris, I.Y., Tsoumakas, G., Vlahavas, I.: A comprehensive study over vlad and product quantization in large-scale image retrieval. IEEE Trans. Multimedia **16**(6), 1713–1728 (2014)
37. Srivastava, A.N., Zane-Ulman, B.: Discovering recurring anomalies in text reports regarding complex space systems. In: Aerospace Conference, pp. 3853–3862. IEEE (2005)
38. Tomás, J.T., Spolaôr, N., Cherman, E.A., Monard, M.C.: A framework to generate synthetic multi-label datasets. Electron. Notes Theoret. Comput. Sci. **302**, 155–176 (2014)
39. Tsoumakas, G., Katakis, I.: Multi-label classification: An overview. Int. J. Data Warehouse. Min. **3**(3), 1–13 (2007)
40. Tsoumakas, G., Katakis, I., Vlahavas, I.: Effective and efficient multilabel classification in domains with large number of labels. In: Proceedings of ECML/PKDD Workshop on Mining Multidimensional Data, MMD'08, pp. 30–44 (2008)
41. Tsoumakas, G., Katakis, I., Vlahavas, I.: Mining multi-label data. In: Data Mining and Knowledge Discovery Handbook, pp. 667–685. Springer (2010)
42. Tsoumakas, G., Vlahavas, I.: Random k-Labelsets: an ensemble method for multilabel classification. In: Proceedings of 18th European Conference on Machine Learning, ECML'07, vol. 4701, pp. 406–417. Springer (2007)
43. Tsoumakas, G., Xioufis, E.S., Vilcek, J., Vlahavas, I.: MULAN multi-label dataset repository. http://mulan.sourceforge.net/datasets.html

44. Turnbull, D., Barrington, L., Torres, D., Lanckriet, G.: Semantic annotation and retrieval of music and sound effects. IEEE Trans. Audio Speech Lang. Process. **16**(2), 467–476 (2008)
45. Turner, M.D., Chakrabarti, C., Jones, T.B., Xu, J.F., Fox, P.T., Luger, G.F., Laird, A.R., Turner, J.A.: Automated annotation of functional imaging experiments via multi-label classification. Front. Neurosci. **7** (2013)
46. Tzanetakis, G., Cook, P.: Musical genre classification of audio signals. IEEE Trans. Speech Audio Process. **10**(5), 293–302 (2002)
47. Ueda, N., Saito, K.: Parametric mixture models for multi-labeled text. In: Proceedings of 15th Annual Conference on Neural Information Processing Systems, NIPS'02, pp. 721–728 (2002)
48. Wieczorkowska, A., Synak, P., Raś, Z.: Multi-label classification of emotions in music. In: Intelligent Information Processing and Web Mining, AISC, vol. 35, chap. 30, pp. 307–315 (2006)
49. Zhang, M., Zhou, Z.: ML-KNN: a lazy learning approach to multi-label learning. Pattern Recogn. **40**(7), 2038–2048 (2007)

Chapter 4
Transformation-Based Classifiers

Abstract One of the first approaches to accomplish multilabel classification was based on data transformation techniques. These are aimed to produce binary or multiclass datasets from the multilabel original ones, thus allowing the use of traditional classification algorithms to solve the problem. The goal of this chapter is to introduce the most relevant transformation-based MLC methods, as well as to experimentally test the most popular ones. Section 4.1 provides a broad introduction to the chapter contents. The main data transformation approaches are defined in Sect. 4.2; then, several methods based on each approach are described in Sects. 4.3 and 4.4. Four of these methods are experimentally tested in Sect. 4.5. Section 4.6 summarizes the chapter.

4.1 Introduction

Also known as problem transformation methods, the transformation-based multilabel classification algorithms aim to convert the original dataset into one or more simpler datasets that can be delivered to traditional classification algorithms. In a certain way, these methods act as a preprocessing phase, producing new datasets from the original ones, combined with a post-prediction step in charge of uniting the individual votes from each classifier into a joined prediction.

The most influential transformation algorithms are described in the first part of this chapter. All of them follow one of two alternatives, generating binary or multiclass datasets as intermediate representation. Once the datasets have been transformed into simpler ones, the task can rely on standard classifiers and ensembles of classifiers. These two approaches are going to be introduced in the following section and detailed in the further ones, including descriptions of several specific algorithms based on them. Some of them will be experimentally tested in the second part of the chapter.

© Springer International Publishing Switzerland 2016
F. Herrera et al., *Multilabel Classification*,
DOI 10.1007/978-3-319-41111-8_4

4.2   Multilabel Data Transformation Approaches

The learning process to obtain a multilabel classification model is analogous to that of the traditional ones, following a supervised learning path most of the time. The essential difference comes from the need for producing several outputs at once. As many as k different labels, k being the number of elements in the $\mathcal{L}$ set have to be predicted. Each output is of binary nature; therefore, only two possible values are allowed.

When it comes to the development of new multilabel classification algorithms, mainly two approaches are adopted. The first one relies on existing classification algorithms, transforming the original data so it can be adequately processed with them. The second alternative is also founded on the use of already known algorithms, but aims to adapt them to make them able to deal with multilabel data, without internal transformations. This chapter is focused in the former approach.

Transforming an MLD into a simpler representation, suitable for already available classification methods, usually means converting it [2] into a set of binary datasets (BIDs) or into one or more multiclass datasets (MCD). BIDs, MCDs, and MLDs share the fact that a set of predictor variables are available to learn from. The fundamental difference appears in the predicted output. BIDs and MCDs only have one class, which is binary in the former case and can hold more than two possible values in the latter. MLDs, on the contrary, have a set of outputs rather than only one. Each output is a binary value, indicating whether a label is relevant to the corresponding data instance or it is not. This difference is shown in Fig. 4.1, where a tabular representation of a BID (only one binary output), an MCD (only one multivalued output), and an MLD with k binary outputs is depicted. Most classification algorithms are designed to work with datasets having only one output, whether it is binary or multivalued.

Binary dataset

x_1	x_2	...	x_{f-1}	x_f	class
X_{11}	X_{12}	...	X_{1f-1}	X_{1f}	1
X_{21}	X_{22}	...	X_{2f-1}	X_{2f}	0
...					1
X_{n1}	X_{n2}	...	X_{nf-1}	X_{nf}	0

Multiclass dataset

x_1	x_2	...	x_{f-1}	x_f	class
X_{11}	X_{12}	...	X_{1f-1}	X_{1f}	C2
X_{21}	X_{22}	...	X_{2f-1}	X_{2f}	C1
...					C5
X_{n1}	X_{n2}	...	X_{nf-1}	X_{nf}	C3

Multilabel dataset

x_1	x_2	...	x_{f-1}	x_f	y_1	...	y_k
X_{11}	X_{12}	...	X_{1f-1}	X_{1f}	1	0	1
X_{21}	X_{22}	...	X_{2f-1}	X_{2f}	0	1	1
...					1	0	0
X_{n1}	X_{n2}	...	X_{nf-1}	X_{nf}	0	1	0

Fig. 4.1 Tabular representation of a binary dataset, a multiclass dataset, and a multilabel dataset

Beyond some basic transformations, such as the ones mentioned in Sect. 2.4.1 consisting in ignoring the multilabel instances or choosing only one of the labels, most of the published methods aim to transform the multilabel samples into binary or multiclass instances. In the following sections, the main proposals on how to transform an MLD into one or more BIDs/MCDs are detailed. Firstly, the binary-based transformation methods are explained. Then, the multiclass-based ones will be illustrated.

4.3 Binary Classification Based Methods

One of the easiest ways to face multilabel classification consists in taking one label at a time, training individual classifiers for each one of them. Since the labels only can be active (a relevant label) or inactive (a non-relevant label), the result is a binary dataset per label.

Depending on how the binary datasets are combined to train the classifiers, a different model is obtained. The two main approaches are OVO and OVA, discussed below. More complex ensembles of binary classifiers are also further detailed.

4.3.1 OVO Versus OVA Approaches

The OVO and OVA approaches are well-known techniques to face multiclass classification [5] by means of ensembles of binary classifiers. OVA, also known as OVR (*One vs the rest*), is based on the training of an individual classifier for each class against all others. This technique is extended in the multilabel field to take into account that several labels can be relevant at once. Therefore, a binary classifier is trained for each label, as depicted in Fig. 4.2, and their outputs are combined to generate the predicted labelset. In the multilabel field, this approach is named Binary Relevance (BR) [6].

Any existing classification algorithm can be used as underlying classifier. This includes tree-based methods, SVMs, ANNs, and instance-based algorithms. As can be seen, the number of binary classifiers needed to build this solutions is k, the number of labels associated to the MLD. In each BID, only the considered label is taken into account. After training, the test samples are given as input to each binary classifier. The final step consists in joining these individual outputs to form the labelset to be returned as result.

BR is an easy way for implementing an MLC method, relying in a collection of binary classifiers and a simple strategy to combine the outputs. Nonetheless, it also has some inconveniences. The most remarkable is the fact that, by training independent classifiers, it completely dismisses the potential correlations among labels. Additionally, by taking as negative all samples in which the considered label is not relevant, each binary classifier has to deal with extremely imbalanced data.

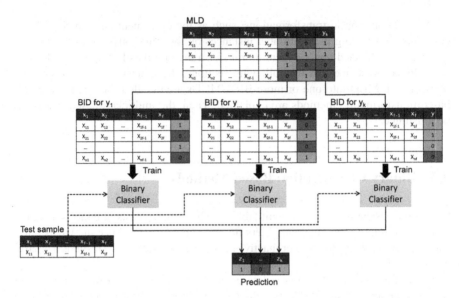

Fig. 4.2 Binary Relevance transformation diagram. The original MLD (*top*) generates as many BIDs as labels there are, using each one to train a binary classifier. When a new test sample arrives, it is given to each individual classifier, joining their predictions to obtain the final labelset

Regarding the OVO approach, in multiclass classification, it is based on the training of classifiers for each pair of classes. This way a specialized model for each pair is obtained, albeit at the cost of a larger collection of classifiers when compared with OVA. In the multilabel field, this idea has been implemented in RPC (*Ranking by Pairwise Comparison*) [7] and CLR (*Calibrated Label Ranking*) [4], among other proposals.

Both algorithms, CLR and RPC, train $k(k-1)/2$ binary classifiers, considering all possible label pairs. Those instances in which only one label of the considered duo is active are included in the transformed binary dataset. The remainder ones, samples where the two labels are active or none of them appear, are disregarded. This way the binary classifier has to learn how to distinguish between the two labels. The outputs of the binary classifiers are used as votes, generating a ranking of labels to be included in the predicted labelset. The CLR algorithm introduces a fictional label, using it to automatically adjust the cut threshold aiming to produce the binary partition of labels from the ranking.

4.3.2 Ensembles of Binary Classifiers

BR is maybe the simplest form of a multilabel ensemble based on binary classifiers. There are as many models as labels, and the strategy to combine their predictions is also simple, just joining the individual outputs to conform the final labelset. RPC and CLR are a little more complex, and there are other proposals also based on ensembles of binary classifiers. The following are some of the most noteworthy:

- **2BR**: This algorithm was proposed [13] aiming to mitigate the most remarkable drawback of BR, which is not taking into consideration label dependencies. 2BR is founded on a stacking strategy, using BR at two levels. The first one learns the BR model, while the second one, taking the outputs of the previous one, learns a meta-model which includes a explicit coefficient for correlated labels.
- **BR+**: The proposal made in [1] has the same goal of 2BR, incorporating label dependency information to increase the multilabel classifier performance. To do so, the binary classifier for each label gains as additional attributes the remainder labels while training. During testing, the sample is processed with a standard BR approach, then the outputs of each classifier are used to enrich the instance input space, and it is introduced in the extended classifiers trained with the additional labels.
- **CC**: It is maybe the best-known multilabel ensemble of binary classifiers. Classifier chains [11] also train k models, as many as labels there are in the MLD, choosing them in random order. The first classifier is trained using only the original input attributes. The first output label is then added as new input attribute, and the new input space is used to train the second classifier, and so on. This way the classifiers are chained, taking into account the possible label dependencies. It is easy to see that different orders of labels can produce disparate results. This is the reason why ECC (*Ensemble of Classifier Chains*) is also proposed, as a set of CC chains with diverse orders and trained with subsets of the available data. Each chain in the ensemble gives a labelset as output which is considered as a set of votes.

The main obstacle of most multilabel ensembles based on binary classifiers is their computational complexity. The processing of MLDs having thousands of labels can be infeasible.

4.4 Multiclass Classification-Based Methods

The other main data transformation path for MLDs relies on the use of multiclass classifiers, instead of binary ones. The basic idea consists in treating each label combination as an unique class identifier, as depicted in Fig. 4.3. This corresponds to the basic LP method proposed in [3] where only a multiclass labelset is generated; thereby, a single multiclass classifier is needed to obtain the predictions. Those are then back-translated to labelsets.

Any multiclass classification algorithm can be used as underlying classifier, thus taking advantage of the accumulated experience in the multiclass field. Since many multiclass classification algorithms operate themselves as transformation methods, producing binary classifiers with the OVA or OVO approach, eventually the multilabel problem could be faced as multiple binary ones depending on the chosen algorithm.

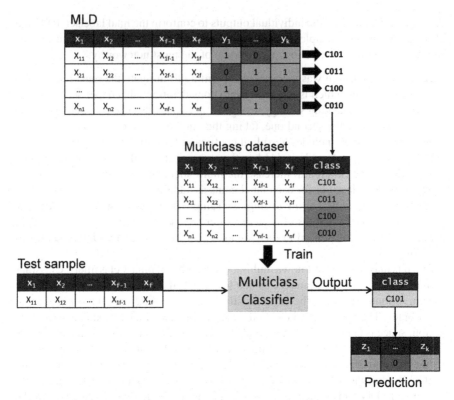

Fig. 4.3 Label Powerset transformation diagram. The labelsets in the original MLD are translated to unique class identifiers and then used to train a multiclass classifier. After processing a new test sample, the output is back-translated to obtain the label predictions

4.4.1   Labelsets and Pruned Labesets

The LP transformation method is even simpler than BR, since there is no need to train several models nor to combine their outputs. However, it also has a major drawback. The number of different label combinations in a large dataset, with a high number of labels and instances, can be incredibly huge.

Theoretically, 2^k different label combinations can exist into an MLD with k labels. In practice, this number is limited by the number of instances in the MLD, because usually $2^k \gg n$. For instance, the delicious dataset has 8.17×10^{295} potential different labelsets, but only 16 105 instances. This circumstance causes that some MLDs have as many distinct labelsets as instances there are in it. Using the LP transformation with such a MLD will produce a multiclass dataset with a different class for each instance; thus, the classifier will face a hard work to train an efficient model.

Unlike BR, the basic LP transformation method implicitly includes in the learning process the hidden relationships among labels. Sets of labels with a certain correlation tend to generate the same or similar label combinations. However, the larger is the number of total labels in the MLD, the higher the likelihood of having a combinatorial explosion that makes this advantage almost useless.

Another obstacle in the use of the LP transformation method is its incompleteness. An LP classifier only can predict label combinations appearing in the training set. These usually represent only a small fraction of the 2^k potential label combinations. On the contrary, an MLC algorithm based on binarization techniques can virtually produce every single labelset.

To avoid the previous problems, a pruning method named PS (*Pruned Sets*) is proposed in [9]. The goal is to focus the classifier onto the key label combinations in the MLD, these which occurs more frequently. For doing so, PS looks for infrequent labelsets and split them into more common combinations. This way the aforementioned combinatorial explosion is alleviated, and the generalization ability of the classifier is improved.

4.4.2 Ensembles of Multiclass Classifiers

Unlike BR, the basic LP transformation method, as well as the PS algorithm, is not an ensemble of classifiers, since only one multiclass classifier is used. Nonetheless, there are several proposals of MLC methods based on ensembles of multiclass classifiers. Some of the most popular are the following. They will be described in greater detail in Chap. 6.

- **EPS**: Introduced in [10] as an extension of the PS algorithm already mentioned before. It trains several independent PS classifiers using a subset of the training data for each one.
- **RAkEL**: Presented in [15], it is a method that generates random subsets of labels, training a multiclass classifier for each subset.
- **HOMER**: Introduced in [14], it is an algorithm that trains several LP classifiers after grouping the training instances into several groups relying in a clustering algorithm.

Although most LP-based methods claim they are able to take label correlations into account, and therefore, they are superior to the BR approach, their effectiveness is largely impacted by the presence of a high number of low-frequent combinations in the MLD.

In addition to MLC methods purely based on the BR and LP transformations, there are some proposals which aim for combining them. It is the case of the ChiDep algorithm [12] that divides the MLC task into the following steps:

1. Firstly, labels with dependencies are identified by applying a χ^2 independence test. Two groups of labels are created, one with the dependent labels and other with the independent ones.

2. The independent labels are processed with a BR classifier, getting individual predictions per label.
3. The dependent labels are processed with a LP classifier, getting join predictions by labelset.
4. The predictions obtained from the two previous steps are merged to produce the final result.

4.5 Data Transformation Methods in Practice

Having known the most popular data transformation-based methods to face multilabel classification, in this part some of them are going to be used and compared aiming to study their behavior. In particular, the following four methods are going to be used:

- **BR**: It is the most basic transformation method based on binarization. It follows the OVA approach.
- **CLR**: A more elaborated binarization method than BR, following the OVO approach.
- **LP**: It is the most basic transformation method based on label combinations. LP is usually compared with BR since it takes label dependency information into account.
- **PS**: A more elaborated labelset method than LP, including a pruning process that alleviates the combinatorial explosion.

These four algorithms have been tested according to the experimental configuration described below. The obtained classification results are presented in Sect. 4.5.2.

4.5.1 Experimental Configuration

The aforementioned data transformation methods needed a binary or multiclass classification algorithm to accomplish their job. The well-known C4.5 [8] tree induction algorithm was used in all cases. Regarding the MLC method configuration, they are always set with their default or recommended parameters.

Our goal is to show how this kind of algorithms behave when used with MLDs having different traits. Therefore, to study BR, CLR, LP, and PS, five heterogeneous MLDs have been selected. These are corel5k, genbase, mediamill, medical, and tmc2007. Their basic traits are the shown in Table 4.1. The remainder characteristics can be found in the tables provided in Chap. 3. In accordance with the *TCS* values shown in Table 4.1, genbase would be the simplest MLD in this group, followed by medical, mediamill, corel5k, and, as the most complex one, tmc2007.

Each dataset was randomly partitioned using a 2×5 fold cross-validation scheme, so each algorithm was ran 10 times over each MLD. To assess the MLC method

Table 4.1 Basic traits of MLDs used in the experimentation

Dataset	n	f	k	Card	Dens	TCS
corel5k	5 000	499	374	3.522	0.009	20.200
genbase	662	1 186	27	1.252	0.046	13.840
mediamill	43 907	120	101	4.376	0.043	18.191
medical	978	1 449	45	1.245	0.028	15.629
tmc2007	28 596	49 060	22	2.158	0.098	21.093

performance, three evaluation metrics have been used, *HLoss* (Hamming loss), *F-measure*, and *SubsetAcc* (Subset accuracy). All of them are example-based metrics, and they were detailed in Chap. 3. The latter one is a very strict performance metric, since it only accounts as correct those predictions in which the full labelsets coincide. Average values were obtained for each measure/MLD across the ten runs. Moreover, training and testing times for each algorithm are also analyzed.

4.5.2 Classification Results

HLoss is a loss metric that sums the number of miss predictions in all the test samples, whether they are false positives or false negatives, and then averages this quantity by the number of instances and labels. Therefore, the lower is *HLoss*, the better is considered the performance of the classifier. The values obtained in this experimentation are represented in Fig. 4.4 and shown in Table 4.2. Best values are highlighted in bold.

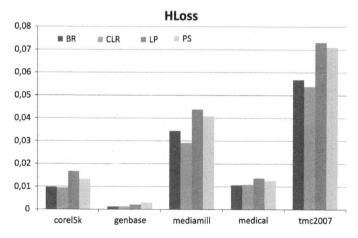

Fig. 4.4 Classification results assessed with Hamming Loss (lower is better)

Table 4.2 Results assessed with *HLoss* (lower is better)

Dataset	BR	CLR	LP	PS
corel5k	0.0098	**0.0095**	0.0167	0.0133
genbase	**0.0012**	0.0014	0.0021	0.0029
mediamill	0.0343	**0.0291**	0.0438	0.0409
medical	**0.0107**	0.0109	0.0137	0.0126
tmc2007	0.0568	**0.0538**	0.0732	0.0711

From these result observations, it can be deducted that, in general, the more complex is the MLD, the worse the MLC methods. According to this evaluation metric, the binarization methods BR and CLR clearly outperformed the ones based on label combinations, gathering all best values.

As defined in Sect. 3.5.1 (see formulation in Chap. 3), *F-measure* is computed as the harmonic mean of *Precision* and *Recall*. This metric is averaged by the number of instances in the test set, but not by the number of labels. Thus, it is not influenced by this factor as is the case of *HLoss*. The *F-measure* values are shown in Fig. 4.5 and Table 4.3.

According to *F-measure*, the corel5k MLD obtained the worst classification results, instead of tmc2007. The best performance corresponds to the CLR and PS algorithms, but a win for BR. CLR and PS can be considered more advanced transformation approaches than LP and BR, so that they achieve best performances fall within the expected.

Lastly, classification results assessed with the *SubsetAcc* metric are provided in Fig. 4.6 and Table 4.4. As mentioned before, this is the strictest evaluation metric, so it is not strange that lower values than with *F-measure* are obtained in all cases.

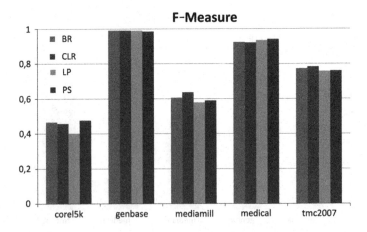

Fig. 4.5 Classification results assessed with *F-measure* (higher is better)

Table 4.3 Results assessed with F-measure (higher is better)

Dataset	BR	CLR	LP	PS
corel5k	0.4654	0.4570	0.3991	**0.4764**
genbase	**0.9910**	0.9906	0.9906	0.9848
mediamill	0.6058	**0.6367**	0.5793	0.5899
medical	0.9239	0.9214	0.9346	**0.9412**
tmc2007	0.7731	**0.7835**	0.7570	0.7604

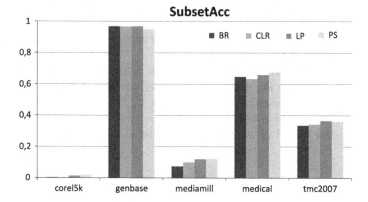

Fig. 4.6 Classification results assessed with *SubsetAcc* (higher is better)

Table 4.4 Results assessed with Subset Accuracy (higher is better)

Dataset	BR	CLR	LP	PS
corel5k	0.0033	0.0016	0.0136	**0.0194**
genbase	**0.9683**	0.9676	0.9676	0.9494
mediamill	0.0740	0.0993	0.1212	**0.1229**
medical	0.6472	0.6340	0.6600	**0.6758**
tmc2007	0.3357	0.3432	**0.3655**	0.3612

Since *SubsetAcc* compares full labelsets, instead of individual labels, the larger is the MLD's set of labels the lower the likelihood of correctly predicting all of them. This is the reason why corel5k and mediamill obtain the worst performances, being the MLDs with more labels out of the five used here.

Another interesting fact can be stated analyzing the best values in Table 4.4. With the only exception of genbase, all the best performances correspond to labelset-based algorithms. This is consistent with the nature of the evaluation metric.

As usual when evaluating an MLC task, that the assessed performance will change depending not only on the selected algorithm, but also on the MLDs traits and chosen evaluation metric must be taken into account.

Transformation methods such as BR and LP are at the heart of several of the methods to be described in the following chapter, since many of them rely on binarization and label powerset techniques. In addition, these two approaches along with PS are the foundation of many ensemble-based methods that will be outlined in Chap. 6. Besides their classification performance, it will be also interesting to get a glimpse of these transformation method running times, since they will have a significant impact in a multitude of other MLC algorithms.

The time spent by each method in training the model for each MLD, measured is seconds, is depicted in Fig. 4.7. Exact values are provided in Table 4.5. As can be observed, binarization-based methods (BR and CLR) need more time to build the classifier when compared to those based on label combinations (LP and PS). Overall PS, which used a pruned set of label combinations, is the best performer regarding

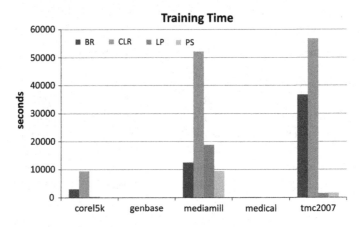

Fig. 4.7 Training time in seconds

Table 4.5 Training time in seconds (lower is better)

Dataset	BR	CLR	LP	PS
corel5k	3 029	9 442	309	**111**
genbase	7	20	**3**	**3**
mediamill	12 468	52 131	18 816	**9 454**
medical	113	203	25	**20**
tmc2007	36 682	56 783	**1 545**	1 632

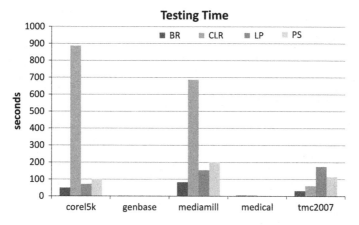

Fig. 4.8 Training time in seconds

Table 4.6 Testing time in seconds (lower is better)

Dataset	BR	CLR	LP	PS
corel5k	**49**	887	72	102
genbase	2	2	2	2
mediamill	**83**	687	155	202
medical	5	7	3	**2**
tmc2007	**31**	61	175	114

this aspect. On the contrary, CLR, with its OVO approach, needs much more time than any other of the tested transformations.

Regarding testing times, the time spent by a classifier to process the test instances predicting the labelset for each one of them, they are shown in Fig. 4.8 and Table 4.6. Once again, the OVO procedure of CLR makes it the slower one. However, once the classifier has been trained, BR is the best performer labeling new instances as can be inferred from the first bar in the plot and the first column in Table 4.6. Therefore, in this case, training time and testing time are inverse criteria.

4.6 Summarizing Comments

Data transformation techniques are a straightforward way to face multilabel classification. In this chapter, the main approaches in this family have been described, and several associated MLC algorithms have been introduced. The diagram in Fig. 4.9 summarizes the data transformation methods described in the previous sections, grouping them according to the strategy they rely on.

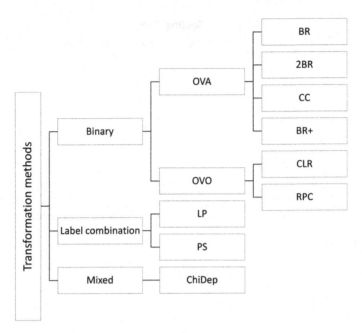

Fig. 4.9 Overview of multilabel data transformation-based methods

Four of the described algorithms, two of them based on binarization techniques and another two on the label powerset principle, have been experimentally tested. Their predictive performance, as well as the time needed to train each model and to use it for prediction, has been analyzed.

References

1. Alvares-Cherman, E., Metz, J., Monard, M.C.: Incorporating label dependency into the binary relevance framework for multi-label classification. Expert Syst. Appl. **39**(2), 1647–1655 (2012)
2. Barot, P., Panchal, M.: Review on various problem transformation methods for classifying multi-label data. Int. J. Data Mining Emerg. Technol. **4**(2), 45–52 (2014)
3. Boutell, M., Luo, J., Shen, X., Brown, C.: Learning multi-label scene classification. Pattern Recogn. **37**(9), 1757–1771 (2004)
4. Fürnkranz, J., Hüllermeier, E., Loza Mencía, E., Brinker, K.: Multilabel classification via calibrated label ranking. Mach. Learn. **73**, 133–153 (2008)
5. Galar, M., Fernández, A., Barrenechea, E., Bustince, H., Herrera, F.: An overview of ensemble methods for binary classifiers in multi-class problems: Experimental study on one-vs-one and one-vs-all schemes. Pattern Recogn. **44**(8), 1761–1776 (2011)
6. Godbole, S., Sarawagi, S.: Discriminative methods for multi-labeled classification. Adv. Knowl. Discov. Data Mining **3056**, 22–30 (2004)
7. Hüllermeier, E., Fürnkranz, J., Cheng, W., Brinker, K.: Label ranking by learning pairwise preferences. Artif. Intell. **172**(16), 1897–1916 (2008)
8. Quinlan, J.R.: C4.5: Programs for Machine Learning (1993)

9. Read, J.: A pruned problem transformation method for multi-label classification. In: Proceedings of New Zealand Computer Science Research Student Conference, NZCSRS'08, pp. 143–150 (2008)
10. Read, J., Pfahringer, B., Holmes, G.: Multi-label classification using ensembles of pruned sets. In: Proceedings of 8th IEEE International Conference on Data Mining, ICDM'08, pp. 995–1000. IEEE (2008)
11. Read, J., Pfahringer, B., Holmes, G., Frank, E.: Classifier chains for multi-label classification. Mach. Learn. **85**, 333–359 (2011)
12. Tenenboim-Chekina, L., Rokach, L., Shapira, B.: Identification of label dependencies for multi-label classification. In: Working Notes of the Second International Workshop on Learning from Multi-Label Data, MLD'10, pp. 53–60 (2010)
13. Tsoumakas, G., Dimou, A., Spyromitros, E., Mezaris, V., Kompatsiaris, I., Vlahavas, I.: Correlation-based pruning of stacked binary relevance models for multi-label learning. In: Proceedings of 1st International Workshop on Learning from Multi-Label Data, MLD'09, pp. 101–116 (2009)
14. Tsoumakas, G., Katakis, I., Vlahavas, I.: Effective and efficient multilabel classification in domains with large number of labels. In: Proceedings of ECML/PKDD Workshop on Mining Multidimensional Data, MMD'08, pp. 30–44 (2008)
15. Tsoumakas, G., Vlahavas, I.: Random k-Labelsets: an ensemble method for multilabel classification. In: Proceedings of 18th European Conference on Machine Learning, ECML'07, vol. 4701, pp. 406–417. Springer (2007)

Chapter 5
Adaptation-Based Classifiers

Abstract While data transformation is a relatively straightforward way to do multilabel classification through traditional classifiers, an alternative approach based on adapting those classifiers to tackle the original multilabeled data also has been also explored. This chapter aims to introduce many of these method adaptations. Most of them rely on traditional algorithms based on the trees, neural networks, instance-based learning, etc. A general overview of them is provided in Sect. 5.1. Then, about thirty different proposals are detailed in Sects. 5.2–5.7, grouped according to the type of model they are founded on. A selection of four algorithms are experimentally tested in Sect. 5.8. Some final remarks are provided in Sect. 5.9.

5.1 Overview

Unlike the problem transformation-based methods presented in the previous chapter, those following the adaptation approach aim to prepare existing classification algorithms to make them able to manage instances with several outputs. The changes that must be introduced in the algorithms can be quite simple or really difficult, depending on the nature of the original method and also the way the existence of several labels is going to be considered.

There are proposals of multilabel classifiers based on trees, neural networks, vector support machines, instance-based learning techniques, and probabilistic methods, among others. About thirty of them are portrayed in the following five sections, each one devoted to one of the just-mentioned categories. An additional section enumerates other types of proposals, such as the ones based on ant colonies or genetic algorithms.

In the second part of this chapter, the behavior of four MLC adaptation-based methods is experimentally checked. These algorithms are representatives of the main four types of models: trees, neural networks, support vector machines, and instance-based learning.

© Springer International Publishing Switzerland 2016   81
F. Herrera et al., *Multilabel Classification*,
DOI 10.1007/978-3-319-41111-8_5

5.2 Tree-Based Methods

Decision trees (DT) are among the easiest to understand classification models. Despite their apparent simplicity, some DT algorithms, such as C4.5 [24], yield a performance that makes them competitive against other learning approaches. The following are some of the multilabel classification methods based on DT proposed on the literature.

5.2.1 Multilabel C4.5, ML-C4.5

The research conducted in [10] aimed to classify genes according to their function, taking into account that the same gene can intervene in several functions. The proposed solution is founded on the well-known C4.5 algorithm, appropriately modified to deal with several labels at once. The two key points of the adapted method are as follows:

- The leaves of the tree contain samples that are associated with a set of labels, instead of to one class only.
- The original entropy measure is adjusted to take into consideration the non-membership probability of the instances to a certain label.

The iterative process to build the tree partitions the instances according to the value of a certain attribute, computing the weighted sum of the entropies of each potential subset of labels. If a sample is associated with more than one label, then it is accumulated several times into this weighted sum.

Since each leaf represents a set of labels, this impacts both the process of labeling each node in the tree and also the further pruning task inherent to C4.5.

5.2.2 Multilabel Alternate Decision Trees, ADTBoost.MH

ADTs (*Alternate Decision Trees*) can be seen as a generalization of traditional decision trees. They were introduced in [15], aiming to propose an alternative way to techniques such as boosting in order to improve the precision of tree-based classifiers.

Taking as reference the ADT idea, and with the goal of extending the AdaBoost.MH model presented in [25], ADTBoost.MH is proposed in [11]. This is an ADT-based model adapted to consider the presence of multiple labels per instance. For doing so, the samples are decomposed following the OVA strategy.

Since ADTBoost.MH trains several models, each one of them being an ADT, it is an MLC method that could be included in the ensemble category as well (see Chap. 6).

5.2.3 Other Tree-Based Proposals

Proposed in [34], ML-Tree is an algorithm to induce classification trees following an hierarchical approach. The original data are seen as an hierarchy, and it is decomposed into several simpler problems using an OVA strategy. An SVM classifier is trained for each case. This way the full dataset at the root node is partitioned into several subsets going to the child nodes. This process is recursively repeated, generating the tree that represents the data hierarchy. As the previous one, the obtained model can be considered as an ensemble of simpler classifiers.

Once the tree has been completed, each leaf provides the set of predicted labels. From this information, a predictive label vector is produced, aiming to model the relationships among labels. Those that frequently appear together are supposed to have some relation level, estimated from the concurrence frequency. This automatic discovery of label correlations is used to improve the predictive performance.

As the previous proposal, LaCova [3] is an algorithm which also relies on a tree to recursively divide the original multilabel data. This divide-and-conquer approach horizontally grows the tree deciding which input feature provides more information and vertically grows it by grouping labels into subnodes.

For each node, a label covariance matrix is built, depending on which the tree will be expanded vertically or horizontally. The dependency among labels in each node is locally assessed, instead of elaborating a global dependency model. When a new vertical split is decided, the instances contained in each new subnode are processed using a BR or LP classifier, subject to the obtained dependency information.

5.3 Neuronal Network-Based Methods

Artificial neural networks (ANNs) in general, and particularly Radial Basis Function Networks (RBFNs), have proven their effectiveness in classification problems, as well as in regression and time series prediction. As a consequence, the adaptation of ANNs to accomplish MLC tasks is a recurrent topic in the literature. The goal of this section is to provide a succinct description of several ANN-based multilabel algorithms.

5.3.1 Multilabel Back-Propagation, BP-MLL

Considered as the first multilabel-adapted ANN, BP-MLL [36] is founded on one of the simplest ANN models, as is the perceptron. The training algorithm chosen for building the model is also well known, back-propagation.

The key aspect in BP-MLL is the introduction of a new error function used while training the ANN, and computed taking into account the fact that each sample contains several labels. Specifically, this new function penalizes the predictions including labels which are not truly relevant for the processed instance.

In BP-MLL, the input layer has as many neurons as input attributes there are in the MLD. The number of units in the output layer is determined by the number of labels considered. The amount of neurons in the hidden layer is also influenced by the number of labels, and they use a sigmoid activation function.

The BP-MLL algorithm produces a label ranking as result while classifying new instances. To decide which labels will be predicted as relevant, a parameter in charge of adjusting the cut threshold has to be set. Configuring this parameter is the main difficult in using BP-MLL.

The proposal made in [19], called I-BP-MLL, overtakes the aforementioned difficulty. To do so, the threshold is automatically adjusted as the committed error is computed during the learning process. This approach produces a custom threshold for each label, instead of a global threshold as BP-MLL does.

5.3.2   Multilabel Radial Basis Function Network, ML-RBF

The algorithm proposed in [37] is an specialized method for designing RBFNs adapted to work with multilabel data. It takes the samples associated with each label and then executes a K-means clustering as many times as labels there are. This way the centers of the RBFs are set. The number of clusters by label is controlled by a α parameter. Depending on the value assigned to α, the number of units in the hidden layer will be equal or larger than the number of labels in the MLD.

ML-RBF uses the SVD (*Singular Value Decomposition*) method to adjust the weights of the connections to the output units, minimizing the squared sum of the computed error in each training iteration. The activation of all neurons is set to 1 and a bias is defined for each label.

In addition to the α parameter, ML-RBF needs another two parameters named σ and μ. The former, usual in RBFNs, controls the width of the unit. ML-RBF computes it by means of an equation in which the distance between each pair of patterns and a scaling factor intervene. The latter sets the scaling factor.

There are two ML-RBF variations, called FPSO-MLRBF and FSVD-MLRBF [2], designed as hybridization of techniques which aim to improve the results produced by the original algorithm. FPSO-MLRBF relies on a fuzzy PSO (*particle swarm optimization*) method with the goal to set the number of units in the hidden layer, as well as to optimize the weights between the hidden and the output layers. FSVD-MLRBF resorts to the use of a fuzzy K-means along with SVD with same goal.

5.3.3 Canonical Correlation Analysis and Extreme Learning Machine, CCA-ELM

Introduced in [20], this algorithm suggests a methodology for adapting an Extreme Learning Machine (ELM) to be able to deal with multilabeled samples. ELMs are a type of ANN with only a hidden layer, characterized by the speed it can learn. This trait makes it ideal to process multilabel datasets.

The proposed method starts using Canonical Correlation Analysis (CCA) in order to detect potential correlations among input attributes and labels. As a result, a new space which combines inputs and labels is generated, being used to train the ELM. Once trained, the ELM can be used to obtain predictions. Those need to be back-translated by applying the inverse transformation to the original solution space, thus obtaining the predicted set of labels.

5.4 Vector Support Machine-Based Methods

Vector Support Machines have been traditionally applied to solve binary classification problems. As other techniques, they have evolved over time, being extended to face other kind of tasks such as multiclass and multilabel classification. In this section, several of the MLC methods based on the SVMs are outlined.

5.4.1 MODEL-x

The target task of the authors in [5] was to label natural landscapes. Each scene can contain several objects at once and, as a consequence, it can be labeled with more than one class. Some scene labels are urban, beach, field, mountain, beach + field, field + mountain, etc.

Since SVMs tend to have a good behavior while dealing with images, the authors selected this kind of model as underlying classifier. The proposed method, called MODEL-x, trains a SVM per label using a novel approach named *cross-training*. This technique takes the samples having several labels as positive cases while training every individual model, instead of negative ones as do the algorithms which consider each label combination as a different class. As many other MLC proposals, MODEL-x can be also considered as an ensemble.

5.4.2 Multilabel SVMs Based on Ranking, Rank-SVM and SCRank-SVM

As the authors of Rank-SVM state in [13], binary classifiers are not the best option when some correlations among labels exist, since full independence between them is assumed by most binary-based methods. To alleviate this problem Rank-SVM, a direct approach based on SVM principles, relies in a new metric to be minimized while training, AHL. This a lineal approximation of Hamming Loss.

Once the SVM-like model has been trained to produce the label ranking, a specific function is used to adjust the cutting threshold from which the predicted labelset, as a subset of the full ranking, is extracted.

Although Rank-SVM takes label correlations into account, thus it theoretically should perform better than pure binary models, experimental tests show that its behavior is similar when working with high-dimensional MLDs.

These two proposals are based on the Rank-SVM, aiming to improve their efficiency and performance. The goal of Rank-CVM [35] is to reduce the computational complexity of Rank-SVM. To do so, they decided to use a CVM (*core vector machine*) instead of an SVM. The analytical solution of CVMs is immediate, and much more efficient than that of SVMs. The result is an MLC classifier with similar predictive performance to Rank-SVM but an order of magnitude more efficient.

The goal of the authors of SCRank-SVM [33] also was to improve the efficiency of Rank-SVM, as well as its performance as MLC classifier. The proposal includes reformulating the calculus of the decision boundary in the underlying SVMs, simplifying some of the existent constraints to maximize the margin. In the process, the authors get rid of one of the usual SVM parameters, reducing the complexity of computations.

5.5 Instance-Based Methods

Unlike the algorithms based on inductive learning, instance-based methods do not build a explicit model trough a training process. They rather work as lazy classification procedures, taking the k nearest neighbors (kNN) when a new data sample arrives. There are several instance-based MLC-adapted proposals, and a few of them are summarized in this section.

5.5.1 Multilabel kNN, ML-kNN

Presented in [38], ML-kNN is the best-known instance-based MLC algorithm. It internally works as a BR classifier, since a separate set of a priori and conditional probabilities are independently computed for each label. Therefore, any potential correlation information among labels is disregarded by ML-kNN. The inner working details of ML-kNN were provided in Sect. 3.4.1.

Owing its simplicity and low computational complexity, ML-kNN is usually included in most experimental studies. It also has been used as foundation for other more elaborated MLC algorithms, such as IBLR-ML.

5.5.2 Instance-Based and Logistic Regression, IBLR-ML

Two similar MLC methods are introduced in [8], both of them based on the afore-mentioned ML-kNN algorithm. The core of the two proposals, named IBLR-ML and IBLR-ML+, use Bayesian techniques to consider the labels associated with nearest neighbors of the new instance as additional characteristics. Using this information the a priori probabilities are computed and a regression equation is obtained.

The main difficulty in these methods comes from the need to adjust an α parameter, in charge of setting the weight to be assigned to the additional attributes while computing the a posteriori probabilities. To accomplish this task an statistical parameter estimation method is used. This is an adaptation process which demands the enumeration of all instances in the MLD.

Although IBLR-ML can achieve better predictive performance than ML-kNN, it also has higher computational demands, including more memory consumption and training time.

5.5.3 Other Instance-Based Classifiers

The kNNc method was proposed in [6]. It works in two stages, combining instance selection techniques with instance-based classification. Firstly, a reduced set of instances is obtained by prototype selection techniques. The aim is to determine the set of labels which are nearest to the ones in the instance to be classified. Then, the full set of samples is used, but limiting the prediction to the labels inferred in the previous step.

BRkNN and LPkNN [27] are MLC classifiers made up combining BR and LP transformation methods with kNN classification techniques, respectively. They first apply the data transformation to the MLD, then use kNN to find the nearest neighbors, and generate the labelset from them.

The proposal in [16] is called BRkNN-new. As its name indicates, it is an improvement based on the BRkNN method just described above. The goal of the authors is to take advantage of label correlation information in order to improve the performance of the original model.

5.6 Probabilistic Methods

Probabilistic techniques, such as Bayesian models, mixture models, or conditional random fields, are commonly used in many MLC methods. Sometimes, these techniques appear only as a part of other algorithms, for instance the Bayesian model used in IBLR-ML, while in other cases they are the cornerstone of the method. This section enumerates some of the proposals in the latter group.

5.6.1 Collectible Multilabel Classifiers, CML and CMLF

The authors of these two algorithms, presented in [17], state that binary MLC methods assume the labels are fully independent, and do not take into account potential correlations among them. This is the reason to propose a model to capture this information, aiming to improve the predictive performance of the classifiers.

In order to get the correlations between labels, a CRF (*Conditional Random Field*) is used. A CRF is a probabilistic discriminative model, commonly used in tasks such as text segmentation. It associates a probability to each label depending on the existent observations, and with these data, the potential dependencies among outputs are modeled. Four different states can exist for each label pair with respect to the considered instance, none of the labels is relevant, both of them are applicable, only the first one is relevant, or only the second one is relevant.

The first proposed model is CML (*Collectible Multilabel*). It holds a correlation parameter for each label pair. The second one is CMLF (*CML with Features*), and it works with three-variable groups such as *attribute-label1-label2*. For each label pair associated with one attribute, CMLF holds a correlation value. Those labels which do not reach a certain frequency are discarded.

5.6.2 Probabilistic Generic Models, PMM1 and PMM2

These two methods were introduced in [31]. They are probabilistic generative models (PMM), whose goal is to automate text document classification by estimating the label probabilities from the terms appearing into the documents.

PMMs assume that each sample in the text MLD is a mixture of characteristic words related to the labels relevant to the instance. The main difference between PMM1 and PMM2 relies on the way the parameters controlling the algorithms are approximated, PMM2 being a more flexible version of PMM1.

5.6.3 Probabilistic Classifier Chains, PCC

PCC (*probabilistic classifier chains*) [9] is an extension of the CC method previously described in Chap. 4. The goal is to use Bayesian methods to optimize the chaining order of the binary classifiers, thus improving the overall classifier performance.

PCC models the dependency among labels computing the joint distribution for all of them, then deciding which is the optimum chaining order. Although this approach achieves better predictive performance than CC, its computational complexity is also much higher.

5.6.4 Bayesian and Tree Naïve Bayes Classifier Chains, BCC and TNBCC

The major drawback of PCC and other similar approaches based on classifier chains is their computational complexity. This is an inconvenience while working with MLDs having large sets of labels, and sometimes, it could be infeasible to use this kind of algorithms.

In [28], the authors propose several extensions to CC which follow a different path, named BCC (*Bayesian classifier chains*). TNBCC (*Tree Naïve BCC*) is an algorithm based on this approach. By using Bayesian networks to model, the dependencies among labels, it reduces the number of chain combinations to consider to finally compose the multilabel classifier.

5.6.5 Conditional Restricted Boltzmann Machines, CRBM

RBMs (*Restricted Boltzmann Machines*) [26] are a proven mechanism when it comes to produce high-level features from the low-level input attributes existent in a dataset. They are a usual component in the process to build deep belief networks, as well as other deep learning structures. An RBM is a probabilistic graphic model, specifically a two-layer graph, one input layer and one hidden layer. The latter is used to model the relationships between the input attributes.

Introduced in [21], the CRBM (*Conditional RBM*) algorithm relies on an RBM to retrieve dependencies among labels. The goal is to build a model able to operate with MLDs containing incomplete sets of labels. For doing so, the labels of each instance are given as input to the RBM, obtaining as output a dependency model capable of predicting missing labels from others presence.

5.7 Other MLC Adaptation-Based Methods

In addition to the more than twenty MLC methods mentioned above, adaptations of trees, ANNs, SVMs, instance-based, and probabilistic classifiers, in the literature can be found a lot more following alternative ways of attacking the problem. The enumerated below are some of them:

- **HG**: It is based on an hypergraph algorithm, each label being one edge, whose goal is to generate a model containing the relationships among labels. The resulting problem is hard to solve from a computational point of view, but the authors in [29] assume certain premises that allow them to accomplish the task as a simplified problem of minimum squares.
- **CLAC**: One of the major obstacles in facing MLC tasks is the usually huge number of label combinations an MLD can hold. This produces a secondary problem, even harder to solve, since the number of data samples sharing the same combination of labels can be almost negligible. These cases are identified in [32] as *disjuncts*, highlighting that if they are ignored, due to their poor representation, the obtained model can loss precision since in large MLDs many disjuncts can exist. The MLC method proposed is a lazy algorithm named CLAC (*Correlated Lazy Associative Classifier*), which increases the relevance of the potential disjuncts by discarding attributes and samples not related to the processed data instance.
- **GACC**: Introduced in [18], the GACC (*Genetic Algorithm for ordering Classifier Chains*) method is proposed as an alternative to the traditional ECC. While the latter randomly generates chains of binary classifiers, the former resorts to the use of a genetic algorithm to optimize the order of the classifiers in the chain. Besides the usual goal of improving classification results, the authors also aim to make the obtained model more easily interpretable.
- **MuLAM**: Although ant colonies based algorithms [12] are mostly used in optimization problems, as ACO (*Ant Colony Optimization*) methods, there are also techniques such as Ant-Miner [23] able to produce classification rules following the same approach. Taking Ant-Miner as foundation, in [7] the authors present a new MLC algorithm named MuLAM (*Multi-Label Ant-Miner*). It takes the multilabel classification problem as an optimization task, but the reported predictive performance seems to be far from the state-of-the-art MLC classifiers.
- **ML-KMPSO**: The proposal made in [22] is an MLC method which combines the kNN and MPSO (*Michigan Particle Swarm Optimization*) algorithms. Firstly, the a priority probabilities of each label are computed. Then, MPSO is used to optimize the selection of nearest neighbors to the considered instance, obtaining a set of expert *particles* which will help to choose the set of labels to be predicted.
- **GEP-MLC**: In [4], the authors present an MLC method founded on discriminant functions optimized by GEP (*Gene Expression Programming* [14]) techniques. GEP is a genetic programming approach specially suitable for regression problems, being used for facing traditional classification as well. GEP-MLC learns one or more discriminant function for each label, thus working as a BR transformation method in which the underlining classifiers are the functions optimized by GEP.

- **MLC-ACL**: Designed as a combination of transformation and adaptation methods, the MLC-ACL algorithm [4] has tree phases. Firstly, the MLD is transformed into a dataset with only one label, following the least frequent criterion. Secondly, a rule-based classification algorithm is applied. Lastly, an iterative process relies on the association rule mining A priori algorithm [1] to detect correlations among labels, transforming the rules obtained in the first step in a multilabel classifier.

5.8 Adapted Methods in Practice

In the previous sections, around thirty adaptation-based multilabel classification algorithms have been introduced. More than half of them can be included in one of the aforementioned four traditional classification approaches. The aim of this section is to test one algorithm of each one of these approaches, comparing the results of a multilabel tree, a multilabel neural network, a multilabel SVM, and a instance-based multilabel method. The selected algorithms are the following:

- **ML-Tree**: It is a recent multilabel classifier that takes into account the potential dependencies among labels, including this information into the tree model. As explained above (see Sect. 5.2.3), it follows an hierarchical approach to generate the tree.
- **BP-MLL**: This method is the first adaptation of an ANN to the multilabel problem, including a specific error function which considers the existence of multiple outputs.
- **Rank-SVM**: The SVM-based multilabel adaptation, as was explained in Sect. 5.4.2, also takes advantage of label correlation information aiming to improve classification performance.
- **BRkNN**: It is a lazy, instance-based multilabel learner which combines the BR transformation with kNN techniques. It is an alternative to the ML-kNN algorithm already used in experiments of previous chapters.

The BP-MLL and BRkNN algorithms are available in the MULAN package [30], while ML-Tree and Rank-SVM can be obtained from the authors' Web site.[1] The former is written in Java and it relies on the WEKA software package, as the latter is implemented in MATLAB.

---

[1] ML-TREE code can be downloaded from Dr. Qingyao Wu's Web page at https://sites.google.com/site/qysite. Rank-SVM code can be downloaded from http://cse.seu.edu.cn/people/zhangml/files/RankSVM.rar.

Table 5.1 Basic traits of MLDs used in the experimentation

Dataset	n	f	k	Card	Dens	TCS
emotions	593	72	6	1.868	0.485	9.364
medical	978	1 449	45	1.245	0.028	15.629
scene	2 407	294	6	1.074	0.179	10.183
slashdot	3 782	1 079	22	1.181	0.054	15.125
yeast	2 417	103	14	4.237	0.303	12.562

5.8.1 Experimental Configuration

The four multilabel classifiers pointed out above have been run over the partitions of five MLDs. These are emotions, medical, scene, slashdot and yeast. Their basic traits are that shown in Table 5.1. The tables in Chap. 3 provide the main characteristics of each one of them. Looking at the *TCS* values in Table 5.1, these MLDs complexity goes from the 9.364 score of emotions to the 15.629 of medical.

As in the previous chapter, each MLD was partitioned using a 2×5 fold cross-validation scheme. The prediction performance is assessed with four evaluation metrics. Since some of the classifiers included in this experimentation produce a ranking as output, in addition to the usual example-based *HLoss* (Hamming loss), three ranking-based evaluation metrics have been selected, *RLoss* (Ranking loss), *OneError*, and *AvgPrecision* (Average precision). The goal is to check how the performance of the classifiers is reported by different evaluation metrics. All of them were detailed in Chap. 3. The reported values in the following section are average values computed from the 10 runs for each MLD/classifier.

5.8.2 Classification Results

As we already know, *HLoss* is a metric to minimize since it is an averaged counter of misclassified labels. In Fig. 5.1, the results for this metric have been represented as $1 - HLoss$, aiming to preserve the perception that a larger area corresponds to a better performance of the classifier. From this plot it is easy to infer that BP-MLL is the worst performer, while the other three seem to show a similar behavior.

The raw *HLoss* values for each MLD/classifier combination are the ones shown in Table 5.2. Best values are highlighted in bold. As can be seen, Rank-SVM obtained the best results for the two most complex MLDs, while BRkNN performed better with the simpler ones.

RLoss is also a loss measure, as *HLoss*, but it is ranking based instead of example based. Therefore, the goal of a classifier must be to minimize this metric. The same previous approach has been used to represent *RLoss* values in the plot shown in

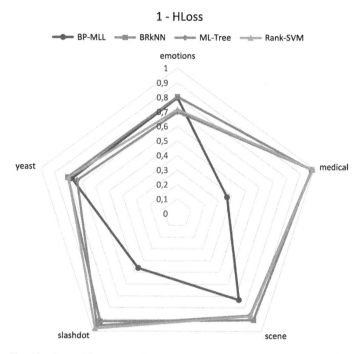

Fig. 5.1 Classification results assessed with Hamming Loss

Table 5.2 Results assessed with *HLoss* (lower is better)

Dataset	BP-MLL	BRkNN	ML-Tree	Rank-SVM
emotions	0.2034	**0.1965**	0.3016	0.2842
medical	0.6347	0.0182	0.0142	**0.0104**
scene	0.2669	**0.0948**	0.1352	0.1288
slashdot	0.5360	0.0834	0.0475	**0.0230**
yeast	0.2265	**0.1967**	0.2641	0.2024

Fig. 5.2. As can be observed, the shape of the classifiers is quite similar to Fig. 5.1, although some larger differences can be detected among the three best performing methods.

Once again that BP-MLL produces the worst classification results, specifically when used with the more complex MLDs, can be stated. In general, the line which corresponds to the Rank-SVM algorithm seems to be the closest to the outer limit, denoting that it is best performed when the results are assessed with this evaluation metric. This perception can be confirmed by examining the values provided in Table 5.3. With the only exception of emotions, the best values always come from the Rank-SVM method.

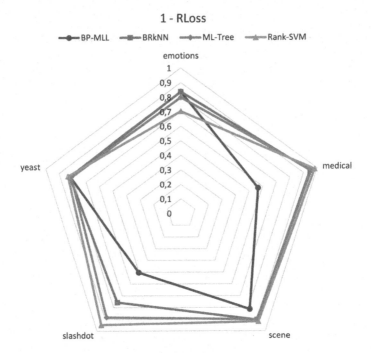

Fig. 5.2 Classification results assessed with Ranking Loss

Table 5.3 Results assessed with *RLoss* (lower is better)

Dataset	BP-MLL	BRkNN	ML-Tree	Rank-SVM
emotions	**0.1625**	0.1629	0.2020	0.2961
medical	0.4340	0.0512	0.0422	**0.0180**
scene	0.1852	0.0955	0.0998	**0.0818**
slashdot	0.4968	0.2430	0.1163	**0.0512**
yeast	0.1854	0.1786	0.1909	**0.1672**

The second ranking-based evaluation metric is *OneError*. It counts how many times the label predicted with more confidence, ranked highest in the output returned by the classifier, is not truly relevant to the instance. As the two previous ones, *OneError* is also a metric to be minimized.

As can be seen in Fig. 5.3, which shows the spider plot for $1 - OneError$, the differences among the classifiers are much clearer with this metric. It can be verified that the three simplest MLDs, `emotions`, `scene` and `yeast`, put fewer obstacles than `medical` and `slashdot` for most of the classifiers. The performance of BP-MLL is particularly poor with these two last MLDs. BRkNN also shows a degradation of its behavior while working with them.

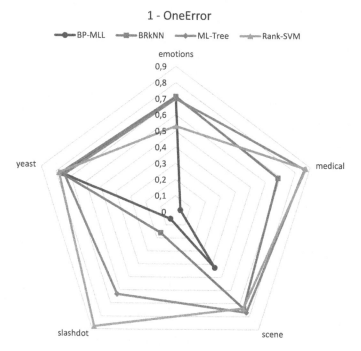

Fig. 5.3 Classification results assessed with One Error

Table 5.4 Results assessed with *OneError* (lower is better)

Dataset	BP-MLL	BRkNN	ML-Tree	Rank-SVM
emotions	**0.2942**	0.2866	0.3010	0.4704
medical	0.9673	0.3160	0.1375	**0.1293**
scene	0.5727	0.2614	**0.2318**	0.2634
slashdot	0.9466	0.8399	0.3691	**0.1221**
yeast	0.2470	0.2317	0.2561	**0.2203**

The raw *OneError* measures are the shown in Table 5.4, with best values highlighted in bold. Rank-SVM achieves best result with the three more complex MLDs. Paradoxically, this algorithm shows its worst performance with the simplest dataset, emotions.

The last metric used to assess the performance in this experimentation is *Avg-Precision*. It measures the number of positions in the ranking that have to be visited before a non-relevant label is found. Therefore, the larger is the value obtained, the better is working the classifier.

In this case, the values produced by the evaluation metric can be plotted without any transformation, as has been done in Fig. 5.4. This plot closely resembles the one in Fig. 5.3; thus, the conclusions drawn from the *OneError* metric can be mostly applied to *AvgPrecision* also.

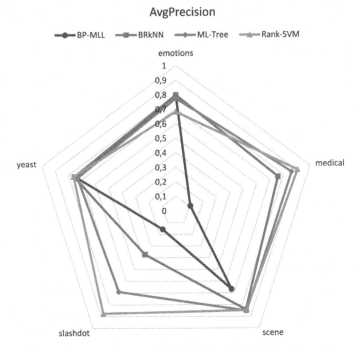

Fig. 5.4 Classification results assessed with *AvgPrecision* (higher is better)

Table 5.5 Results assessed with Average Precision (higher is better)

Dataset	BP-MLL	BRkNN	ML-Tree	Rank-SVM
emotions	0.7971	**0.7979**	0.7737	0.6813
medical	0.1076	0.7594	0.8626	**0.9064**
scene	0.6680	0.8423	**0.8517**	0.8476
slashdot	0.1583	0.3723	0.6892	**0.8746**
yeast	0.7446	0.7583	0.7342	**0.7636**

The set of best values highlighted in Table 5.5 is also very similar to that of Table 5.4. Rank-SVM produces the best results for the most complex MLDs again, while it works quite poor with the simplest one.

The overall conclusion that can be gathered from this experiments is that, from the four tested algorithms, Rank-SVM tends to perform better in most cases. On the contrary, BP-MLL usually produces the worst result with few exceptions. The other two multilabel classifiers, BRkNN and ML-Tree, are in between.

5.9 Summarizing Comments

Traditional pattern classification is a largely studied problem in DM, with dozens of proven algorithms ready to use over binary and multiclass data. Therefore, the adaptation of these methods to work with multilabel data was a clear path from the beginning. In this chapter, several of these MLC adaptation-based algorithms have been introduced.

The diagram in Fig. 5.5 summarizes the algorithm adaptation methods described in the previous sections, grouping them according to the type of base algorithm they are derived from. The first four branches, counting from left to right, correspond to the main four traditional classification approaches, trees, neural networks, SVMs, and instance-based learners.

Four of the enumerated methods, BP-MLL, BRkNN, ML-Tree, and Rank-SVM, have been used in a limited experimentation processing five MLDs. According to the assessment made by four distinct evaluation metrics, the latter shows the best predictive performance, while the former is the worst one.

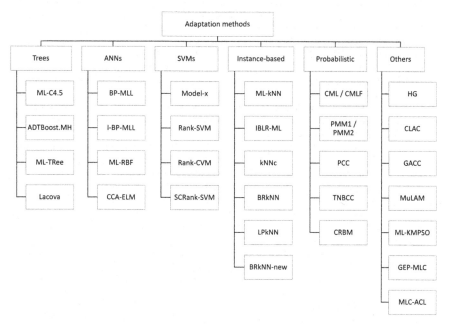

Fig. 5.5 Overview of multilabel algorithm adaptation-based methods

References

1. Agrawal, R., Srikant, R., et al.: Fast algorithms for mining association rules. In: Proceedings of the 20th International Conference on Very Large Data Bases, VLDB'94, pp. 487–499. Morgan Kaufmann (1994)
2. Agrawal, J., Agrawal, S., Kaur, S., Sharma, S.: An Investigation of fuzzy PSO and fuzzy SVD based RBF neural network for multi-label classification. In: Proceedings of the 3rd International Conference on Soft Computing for Problem Solving, SocProS'13, vol. 1, pp. 677–687. Springer (2014)
3. Al-Otaibi, R., Kull, M., Flach, P.: Lacova: a tree-based multi-label classifier using label covariance as splitting criterion. In: Proceedings of the 13th International Conference on Machine Learning and Applications, ICMLA'14, pp. 74–79. IEEE (2014)
4. Ávila, J., Gibaja, E., Ventura, S.: Multi-label classification with gene expression programming. In: Proceedings of the 4th International Conference on Hybrid Artificial Intelligence Systems, HAIS'09, pp. 629–637. Springer (2009)
5. Boutell, M., Luo, J., Shen, X., Brown, C.: Learning multi-label scene classification. Pattern Recogn. **37**(9), 1757–1771 (2004)
6. Calvo-Zaragoza, J., Valero-Mas, J.J., Rico-Juan, J.R.: Improving knn multi-label classification in prototype selection scenarios using class proposals. Pattern Recogn. **48**(5), 1608–1622 (2015)
7. Chan, A., Freitas, A.A.: A new ant colony algorithm for multi-label classification with applications in bioinfomatics. In: Proceedings of the 8th Annual Conference on Genetic and Evolutionary Computation, GECCO'06, pp. 27–34. ACM Press (2006)
8. Cheng, W., Hüllermeier, E.: Combining instance-based learning and logistic regression for multilabel classification. Mach. Learn. **76**(2–3), 211–225 (2009)
9. Cheng, W., Hüllermeier, E., Dembczynski, K.J.: Bayes optimal multilabel classification via probabilistic classifier chains. In: Proceedings of the 27th International Conference on Machine Learning, ICML'10, pp. 279–286 (2010)
10. Clare, A., King, R.D.: Knowledge discovery in multi-label phenotype data. In: Proceedings of the 5th European Conference Principles on Data Mining and Knowledge Discovery, PKDD'01, vol. 2168, pp. 42–53. Springer (2001)
11. De Comité, F., Gilleron, R., Tommasi, M.: Learning multi-label alternating decision trees from texts and data. In: Proceedings of the 3rd International Conference on Machine Learning and Data Mining in Pattern Recognition, MLDM'03, vol. 2734, pp. 35–49. Springer (2003)
12. Dorigo, M., Stützle, T.: Ant Colony Optimization. The MIT Press (2004)
13. Elisseeff, A., Weston, J.: A kernel method for multi-labelled classification. In: Advances in Neural Information Processing Systems, vol. 14, pp. 681–687. MIT Press (2001)
14. Ferreira, C.: Gene expression programming in problem solving. In: Soft Computing and Industry, pp. 635–653. Springer (2002)
15. Freund, Y., Mason, L.: The alternating decision tree learning algorithm. In: Proceedings of the 16th International Conference on Machine Learning, ICML'99, pp. 124–133 (1999)
16. Genga, X., Tanga, Y., Zhua, Y., Chengb, G.: An improved multi-label classification algorithm BRkNN. J. Inf. Comput. Sci. **11**(16), 5927–5936 (2014)
17. Ghamrawi, N., McCallum, A.: Collective multi-label classification. In: Proceedings of the 14th ACM International Conference on Information and Knowledge Management, CIKM'05, pp. 195–200. ACM (2005)
18. Gonçalves, E.C., Plastino, A., Freitas, A.A.: A genetic algorithm for optimizing the label ordering in multi-label classifier chains. In: Proceedings of the 25th IEEE International Conference on Tools with Artificial Intelligence, ICTAI'13, pp. 469–476. IEEE (2013)
19. Grodzicki, R., Mańdziuk, J., Wang, L.: Improved multilabel classification with neural networks. In: Proceedings of the 10th International Conference on Parallel Problem Solving from Nature, PPSN X, pp. 409–416. Springer (2008)

20. Kongsorot, Y., Horata, P.: Multi-label classification with extreme learning machine. In: Proceedings of the 6th International Conference on Knowledge and Smart Technology, KST'14, pp. 81–86. IEEE (2014)
21. Li, X., Zhao, F., Guo, Y.: Conditional restricted boltzmann machines for multi-label learning with incomplete labels. In: Proceedings of the 18th International Conference on Artificial Intelligence and Statistics, AISTATS'15, pp. 635–643 (2015)
22. Liang, Q., Wang, Z., Fan, Y., Liu, C., Yan, X., Hu, C., Yao, H.: Multi-label classification based on particle swarm algorithm. In: Proceedings of the 9th IEEE International Conference on Mobile Ad-hoc and Sensor Networks, MSN'13, pp. 421–424. IEEE (2013)
23. Parpinelli, R.S., Lopes, H.S., Freitas, A.A.: Data mining with an ant colony optimization algorithm. IEEE Trans. Evol. Comput. 6(4), 321–332 (2002)
24. Quinlan, J.R.: C4.5: Programs for machine learning (1993)
25. Schapire, R.E., Singer, Y.: Improved boosting algorithms using confidence-rated predictions. Mach. Learn. 37(3), 297–336 (1999)
26. Smolensky, P.: Information processing in dynamical systems: foundations of harmony theory (1986)
27. Spyromitros, E., Tsoumakas, G., Vlahavas, I.: An empirical study of lazy multilabel classification algorithms. In: Artificial Intelligence: Theories, Models and Applications, pp. 401–406. Springer (2008)
28. Sucar, L.E., Bielza, C., Morales, E.F., Hernandez-Leal, P., Zaragoza, J.H., Larrañaga, P.: Multi-label classification with bayesian network-based chain classifiers. Pattern Recogn. Lett. 41, 14–22 (2014)
29. Sun, L., Ji, S., Ye, J.: Hypergraph spectral learning for multi-label classification. In: Proceedings of the 14th International Conference on Knowledge Discovery and Data Mining, ACM SIGKDD'08, pp. 668–676. ACM (2008)
30. Tsoumakas, G., Xioufis, E.S., Vilcek, J., Vlahavas, I.: MULAN: a java library for multi-label learning. J. Mach. Learn. Res. 12, 2411–2414 (2011)
31. Ueda, N., Saito, K.: Parametric mixture models for multi-labeled text. In: Proceedings of the 15th Annual Conference on Neural Information Processing Systems, NIPS'02, pp. 721–728 (2002)
32. Veloso, A., Meira Jr, W., Gonçalves, M., Zaki, M.: Multi-label lazy associative classification. In: Proceedings of the 11th European Conference on Principles and Practice of Knowledge Discovery in Databases, PKDD'07, pp. 605–612. Springer (2007)
33. Wang, J., Feng, J., Sun, X., Chen, S., Chen, B.: Simplified constraints Rank-SVM for multi-label classification. In: Proceedings of the 6th Chinese Conference on Pattern Recognition, CCPR'14, pp. 229–236. Springer (2014)
34. Wu, Q., Ye, Y., Zhang, H., Chow, T.W., Ho, S.S.: Ml-tree: a tree-structure-based approach to multilabel learning. IEEE Trans. Neural Networks Learn. Syst. 26(3), 430–443 (2015)
35. Xu, J.: Fast multi-label core vector machine. Pattern Recogn. 46(3), 885–898 (2013)
36. Zhang, M.: Multilabel neural networks with applications to functional genomics and text categorization. IEEE Trans. Knowl. Data Eng. 18(10), 1338–1351 (2006)
37. Zhang, M.: Ml-rbf : RBF neural networks for multi-label learning. Neural Process. Lett. 29, 61–74 (2009)
38. Zhang, M., Zhou, Z.: ML-KNN: a lazy learning approach to multi-label learning. Pattern Recogn. 40(7), 2038–2048 (2007)

Chapter 6
Ensemble-Based Classifiers

Abstract Classification methods founded on training several models with a certain heterogeneity degree, and then aggregating their predictions according to a particular strategy tends to be a very effective solution. Ensembles have been also used to tackle some specific obstacles, such as imbalanced class distribution. The goal in this chapter is to present several multilabel ensemble-based solutions. Section 6.1 introduces this approach. Ensembles of binary classifiers are described in Sect. 6.2, while those based on multiclass methods are outlined in Sect. 6.3. Other kinds of ensembles will be briefly portrayed in Sect. 6.4. Some of these solutions are experimentally tested in Sect. 6.5, analyzing their predictive performance and running time. Lastly, Sect. 6.6 summarizes the chapter.

6.1 Introduction

The design of ensembles of classifiers, whether they are founded on binary, multiclass, or heterogeneous classification models, is one of the most common approaches to face multilabel classification tasks. The use of sets of classifiers, along with a strategy to join their individual predictions, has proven to be very effective in traditional classification, so it was another clear path to explore for classifying multilabel data.

Some of the transformation-based and adaptation-based methods discussed in the previous chapters are ensembles by themselves. For instance, the basic BR transformation could be considered as a simple ensemble of binary classifiers with a very straightforward strategy to fuse their individual predictions.

In addition to the simple ensembles mentioned in the previous chapters, in this one a review of more advanced ensembles based on binary classifiers, multiclass classifiers, and other kinds of combinations is provided. Almost a dozen proposals found in the literature will be laid out. Four of them will be experimentally tested and compared.

© Springer International Publishing Switzerland 2016
F. Herrera et al., *Multilabel Classification*,
DOI 10.1007/978-3-319-41111-8_6

6.2 Ensembles of Binary Classifiers

The basic BR data transformation that trains an independent binary classifier for each label and then pastes the predictions to produce the predicted labelset (see Sect. 4.3) is maybe the simplest MLC ensemble. There are many other ensembles also based on binary classifiers, sometimes following an OVA strategy and others resorting to OVO techniques.

In Sect. 4.3.2, once the BR basic transformation was described, some simple ensembles based on binary classifiers, such as 2BR [1], BR+ [2] and CC [3], were also introduced. The aim of this section is to complete the overview of MLC methods based on ensembles of binary classifiers.

6.2.1 Ensemble of Classifier Chains, ECC

One of the weaknesses of CC, already mentioned in Chap. 4, is the fact that the classifiers that make the chain are chosen in a specific order. Since the layout of the chain influences the information given to each classifier in it, any change in the order can also have a great impact in the final results. This is the reason why the authors propose in the same paper [3] the ECC (*Ensemble of Classifier Chains*) method, an ensemble of CC classifiers using a different sorting of binary models in each one of them.

Since CC is an ensemble by itself, ECC can be seen as an ensemble of ensembles. If all training data were used in each CC unit, building the full ECC model would take a very long time for large MLDs. Therefore, only a random subset of the training samples are used in each CC, and the order of labels in each of them is also randomly established. Although CC is inherently a sequential algorithm, since each binary classifier needs the outputs of the previous ones in the chain, ECC can be easily parallelized.

Classifying a new instance with ECC implies going through all the CC classifiers, obtaining a set of predictions for each label. Those are taken as votes for each label, producing a label ranking. Lastly, a threshold is applied to obtain the final multilabel prediction.

6.2.2 Ranking by Pairwise Comparison, RPC

Unlike CC/ECC, the approach followed by RPC (*Ranking by Pairwise Comparison*) [4] is OVO instead of OVA. This means that a binary model is made for each pair of labels, thus increasing the overall number of classifiers to train to $k(k-1)/2$. Linear perceptrons were used as base classifiers for each label pair. Each classifier determines whether a certain label is above or below the other, learning them as preferences for each data sample. Joining the results of all these predictors, an overall label ranking is generated.

The authors also propose in the same paper an improvement of RPC named CMLPC (*Calibrated Pairwise Multilabel Perceptron*). This algorithm is based on a previous method called MLPC, which relies on a perceptron to make pairwise comparisons among labels.

6.2.3 Calibrated Label Ranking, CLR

CLR (*Calibrated Label Ranking*) is an ensemble of binary classifiers proposed in [5]. It is an extension of RPC; hence, it also follows the OVO approach, learning to differentiate between relevance of label pairs.

In addition to the real labels defined in each MLD, CLR introduces in the process a virtual label. It is taken as a reference point which aims to calibrate the final classifier, separating relevant labels of non-relevant ones. Those labels appearing in the ranking above the fictional label will be in the final prediction, while the others will not. This way CLR is a full MLC solution, able to produce a bipartition with the predicted labelset. By comparison, RPC only produces a label ranking, to which some threshold has to be applied to produce the final prediction.

6.3 Ensembles of Multiclass Classifiers

The second main data transformation approach is LP, and it also served as the foundation of several MLC ensemble methods. LP by itself transforms any MLD in a multiclass dataset, so the problem can be faced with only one classifier and there is no need for an ensemble, as in BR. However, the basic LP approach has to deal with other kind of problems, such as the huge number of potential combinations and, therefore, the scarcity of samples representing each produced class.

In order to alleviate the combinatorial explosion when LP is applied to MLDs having large sets of labels, the PS method (see Sect. 4.4.1) proposes a pruning strategy that gets rid of infrequent combinations. Most MLC ensembles based on multiclass classifiers follow a similar approach, getting subsets of labels to reduce the number of obtained combinations.

6.3.1 Ensemble of Pruned Sets, EPS

The *Ensemble of Pruned Sets* [6] method is based on the PS algorithm described in Sect. 4.4.1. It trains m (a user-definable parameter) independent PS classifiers, using a random subset of the training data, typically a 63 % of the available samples, for each one.

The most remarkable novelty of EPS is the introduction of a voting system that allows the prediction of new label combinations, despite they not appearing in the training data. The predictions of each PS classifier are combined, and a threshold is applied to decide which labels will be relevant in the final labelset. Therefore, EPS achieves the completeness that LP and PS lack.

6.3.2 Random k-Labelsets, RAkEL

Random k-Labelsets [7] is a method that generates random subsets of labels, training a multiclass classifier for each subset. This way the problems described in Sect. 4.4.1 for the LP approach are mostly avoided, facing less potential combinations while retaining label correlation information.

RAkEL takes two essential parameters, m and k. The former sets the number of classifiers to train and the latter the length of the labelsets to be generated. With $k = 1$ and $m = |\mathcal{L}|$, RAkEL works as the BR transformation. On the opposite side, with $m = 1$ and $k = |\mathcal{L}|$ the obtained model will be equivalent to LP. Intermediate values for both parameters are the interesting ones, producing a set of votes for each label that produces a label ranking.

6.3.3 Hierarchy of Multilabel Classifiers, HOMER

Introduced in [8], *Hierarchy Of Multilabel classifiERs* is an algorithm designed to deal with MLDs having a large number of labels. The method trains an LP classifier with the available instances and then separates them into several groups relying in a clustering algorithm. Each group, with a subset of the labels, produces a new, more specialized classifier.

The previous step is iteratively repeated, so that each group is divided into several smaller subgroups. The number of iterations in this process is a user-defined parameter given as input to the algorithm. When a test sample arrives, it traverses the hierarchy from its root, where the most general classifier does its work, to the leaves, going through the intermediate nodes activated by the prediction given by the immediate higher level.

6.4 Other Ensembles

In addition to the ensembles purely based on binary or multiclass classifiers, with different variations of the BR and LP data transformations as has been shown, in the literature can be found other MLC ensemble strategies. Several of them are depicted below:

- **CDE**: The authors of the ChiDep algorithm [9], described in Sect. 4.4.2, also proposed CDE (*ChiDep Ensemble*), an MLC ensemble based on ChiDep classifiers. The method starts by generating a large set of random label combinations; then, the χ^2 test is computed for each of them. A parameter m establishes how many of these labelsets will be used to train the individual classifiers in the ensemble, each one following the ChiDep approach.

- **RF-PCT**: This method, introduced in [10], relies in multiobjective decision trees as base classifier. Each tree is able to predict several outputs at once, being a variant of PCTs (*Predictive Clustering Trees*). The ensemble is built following the random forest strategy, hence the RF-PCT (*Random Forest of Predictive Clustering Trees*) denomination. As usual in random forest, bagging techniques are used to boost diversity among the obtained classifiers, using different subsets of the input attributes for each tree.

- **EML**: Unlike other MLC ensembles, the EML (*Ensemble of Multilabel Learners*) proposal [11] relies in an heterogeneous set of MLC classifiers, instead of an homogeneous group of binary, multiclass, or other kinds of classifiers. MLC uses the training data to prepare five different models, such as ECC, ML-kNN, IBLR-ML, RAkEL, and CLR. Then, their outputs are combined using different voting strategies. The main drawback of EML is its computational complexity, since several of its underlying classifiers, such as ECC, RAkEL, and CLR, are ensembles by themselves.

- **CT**: Although proposals such as ECC and BCC, both of them based on classifier chains, achieve a good predictive performance, the process to discover the best structure for the chains is time-consuming. The method proposed in [12], named CT (*Classifier Trellis*), sets a trellis as a priori structure for the ensemble, instead of infer it from the data. Over this fixed structure, the order of the labels is adjusted relying on a simple heuristic which evaluates the frequency between label pairs. The goal was to improve the scalability of chain-based methods while maintaining their performance.

Taking some of the aforementioned ensembles as a foundation, some adjustments in the voting process have been also proposed. It is the case of DLVM (*Dual Layer Voting Method*) [13] and QCLR (*QWeighted CLR*) [14], both aiming to improve classification results through a better strategy to fuse the individual predictions. Likewise, there are some studies focused on methods to improve the setting of the threshold applied to label rankings, such as published in [15].

6.5 Ensemble Methods in Practice

Once most of the multilabel ensemble methods currently available have been introduced, the goal in this section is to put in practice some of them. The following four, maybe the most popular ones, have been selected. All of them are available in the MULAN or MEKA software packages described in Chap. 9.

Table 6.1 Basic traits of MLDs used in the experimentation

Dataset	n	f	k	Card	Dens	TCS
corel16k	13 766	500	153	2.859	0.019	19.722
reuters	6 000	500	103	1.462	0.014	17.548
scene	2 407	294	6	1.074	0.179	10.183
slashdot	3 782	1 079	22	1.181	0.054	15.125
yeast	2 417	103	14	4.237	0.303	12.562

- **ECC**: It is one of the first binary-based multilabel ensembles that takes into account label dependencies, and it has been the foundation for some others.
- **EPS**: This labelset-based ensemble, founded on the PS transformation method, is able to predict label combinations that have not been seen in the training data.
- **RAkEL**: A classical ensemble algorithm to face multilabel classification using random subsets of labels.
- **HOMER**: It is a multilabel ensemble specifically designed to work with MLDs having large sets of labels.

These four multilabel ensembles have been tested using the experimental configuration explained in the following section. Obtained results are presented and discussed in Sects. 6.5.2 and 6.5.3.

6.5.1 Experimental Configuration

The four multilabel ensemble methods selected for this experimentation are transformation-based, so they need an underlying classifier to accomplish each binary or multiclass classification task. As we did to test the multilabel transformation-based methods described in Chap. 4, the C4.5 [16] tree induction algorithm has been chosen for this job. Default or recommended values were used for all the methods parameters.

Each multilabel ensemble method was ran using five MLDs. These are corel16k, reuters,[1] scene, slashdot, and yeast. The main traits of these datasets can be found in the tables provided in Chap. 3. The basic ones are provided in Table 6.1. According to their *TCS* values (see full list in Table 3.5), scene would be the simplest MLD in this group, while corel16k would be the most complex.

As in the previous chapters, each MLD was partitioned with a 2 × 5 folds cross-validation scheme. In order to assess classification performance of each ensemble, three example-based metrics have been used, *HLoss* (Hamming loss), *F-measure* and *SubsetAcc* (Subset accuracy). The details about these metrics can be also found in Chap. 3. Average values were obtained for each metric across the runs for each MLD/algorithm.

---

[1] The version of this MLD having the 500 most relevant features selected was used.

In addition to predictive performance metrics, the training and testing times for each method have been also obtained. This will allow to compare the ensembles from another point of view, the time spent on generating them and the speed with they are able to provide predictions for new data.

6.5.2 Classification Results

The analysis of classification results starts with the *HLoss* metric, as usual. The *HLoss* values for each ensemble and MLD have been plotted as bars in Fig. 6.1. Remember that for *HLoss*, being it a measure to minimize, the taller is the bar and the worse is performing the classifier.

At first sight, it seems that the worst results correspond to `scene` and `yeast`, the two least complex MLDs, while with `corel16k`, which has the highest complexity (according to its *TCS* value), apparently all the ensembles behave quite well. However, it must be taken into account that *HLoss* is also heavily influenced by the number of labels, and `corel16k` has many more than `scene` or `yeast`. That is the reason behind these results. Nonetheless, *HLoss* values belonging to the same MLD but different classifier can be compared. This allows to infer that EPS is always performing better than the other ensembles. On the contrary, the results for ECC put it as the weakest ensemble.

The *HLoss* exact values are those shown in Table 6.2. As always, best values have been highlighted in bold. That EPS achieves the best results can be confirmed, with the only exception of `slashdot` by a small margin to RAkEL.

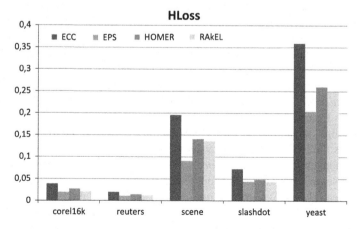

Fig. 6.1 Classification results assessed with Hamming Loss (lower is better)

Table 6.2 Results assessed with Hamming Loss (lower is better)

Dataset	ECC	EPS	HOMER	RAkEL
corel16k	0.0387	**0.0196**	0.0271	0.0206
reuters	0.0198	**0.0112**	0.0146	0.0121
scene	0.1958	**0.0908**	0.1407	0.1363
slashdot	0.0728	0.0440	0.0483	**0.0434**
yeast	0.3594	**0.2042**	0.2601	0.2494

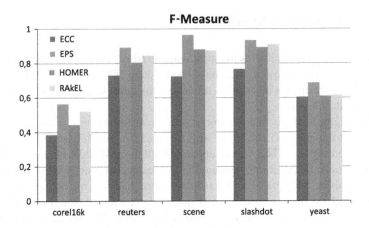

Fig. 6.2 Classification results assessed with F-measure (higher is better)

F-measure is an example-based evaluation metric that provides a good overall indicator of the classifier behavior, as it is the harmonic mean of *Precision* and *Recall*. Classification results assessed with this metric are shown in Fig. 6.2. Looking at the overall results by dataset, it can be seen that the best performance is achieved with scene, while the worst behavior is observed while working with corel16k. The former is the simplest MLD and the latter the most complex according to the *TCS* characterization metric. Regarding the classifiers, that EPS is again the top performer can be deducted. On the contrary, ECC seems to be the one with poorest results. So, the conclusions would be the same drawn from *HLoss* values.

Exact *F-measure* values are shown in Table 6.3. As shown, all highlighted best values come from the EPS ensemble. RAkEL and HOMER perform at a similar level, with slight advantage for one or the other depending on the MLD. In general, ECC obtains the lowest *F-measure* values. Therefore, in agreement with this limited experimentation results, the *F-measure* metric indicates that EPS would be the preferred ensemble from the set of four tested algorithms.

As we already know, *SubsetAcc* is one of the strictest multilabel evaluation metrics. It only accounts as correct predictions those in which the full labelset coincides with the true one. For that reason, *SubsetAcc* values tend to be quite low when compared with other metrics, such as *F-measure*. This is a fact easily inferrable from Fig. 6.3,

Table 6.3 Results assessed with F-measure (higher is better)

Dataset	ECC	EPS	HOMER	RAkEL
corel16k	0.3854	**0.5638**	0.4440	0.5202
reuters	0.7311	**0.8931**	0.8048	0.8452
scene	0.7247	**0.9656**	0.8808	0.8756
slashdot	0.7668	**0.9347**	0.8940	0.9096
yeast	0.6041	**0.6875**	0.6097	0.6149

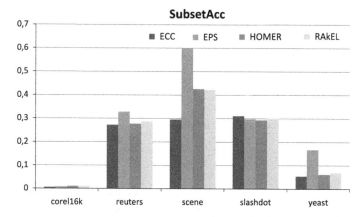

Fig. 6.3 Classification results assessed with Subset Accuracy (higher is better)

Table 6.4 Results assessed with Subset Accuracy (higher is better)

Dataset	ECC	EPS	HOMER	RAkEL
corel16k	0.0059	0.0084	0.0102	**0.0112**
reuters	0.2720	**0.3284**	0.2770	0.2868
scene	0.2952	**0.6006**	0.4260	0.4225
slashdot	**0.3105**	0.2970	0.2931	0.3012
yeast	0.0536	**0.1676**	0.0614	0.0689

where the values for corel16k are almost negligible. The results for EPS stand out with reuters, scene and yeast. The values with slashdot are quite similar for the four methods.

By examining the raw *SubsetAcc* values, provided in Table 6.4, that the results are not as conclusive as they were with *F-measure* can be understood. EPS is still the ensemble that more best values achieves, but ECC works better with slashdot and RAkEL with corel16k. It must be taken into account that, by the nature of the *SubsetAcc* metric, the number of labels on each MLD has a remarkable impact in these results.

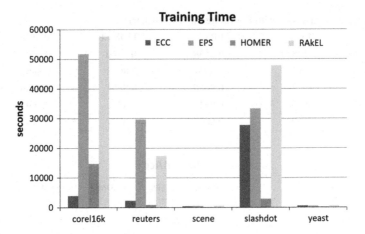

Fig. 6.4 Training time in seconds

Table 6.5 Training time in seconds (lower is better)

Dataset	ECC	EPS	HOMER	RAkEL
corel16k	**3 919**	51 761	14 751	57 632
reuters	2 269	29 660	**850**	17 402
scene	387	514	**85**	545
slashdot	27 739	33 363	**2 851**	47 874
yeast	545	463	**99**	543

6.5.3 Training and Testing Times

Ensemble methods usually demand much more training time than standard classification algorithms, since they have to train several models. Their structure also
influences the time spent in elaborating the predictions for new data samples, going
through each individual classifier and putting together the votes to obtain a united
output. For these reasons, it would be interesting to analyze training and testing times
for each case.

Figure 6.4 shows training times for each MLD and ensemble, measured in seconds.
The training time for the datasets with fewer data samples, scene and yeast, are
almost imperceptible in this graphic representation. For the other tree MLDs, RAkEL
and EPS emerge as the methods that needed more time to be trained. On the other
side, HOMER seems to be the most efficient algorithm. This fact can be verified by
looking at the exact values reported in Table 6.5. It can be observed that for some
MLDs, such as slashdot, HOMER only needs a fraction of the time used by the
other ensembles. This is consistent with the design of HOMER, whose authors aimed
to build an ensemble able to process large MLDs.

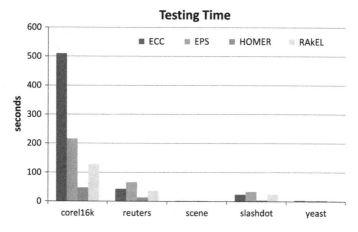

Fig. 6.5 Testing time in seconds

Table 6.6 Testing time in seconds (lower is better)

Dataset	ECC	EPS	HOMER	RAkEL
corel16k	510	216	**47**	128
reuters	43	66	**13**	36
scene	2	2	2	2
slashdot	24	33	**4**	24
yeast	3	3	3	**2**

Regarding the testing times, they are much lower than the training ones as would be expected. In fact, those corresponding to some MLDs are almost undetectable in Fig. 6.5.

Table 6.6 provides the raw values for each MLD/classifier combination. The conclusion that can be drawn from these numbers is that HOMER is the quickest ensemble, in both training and testing. However, it does not achieve best classification results never in the succinct experimentation conducted in this chapter. On the other hand, EPS always produces good classification performance, but its training and testing times are among the worst ones. In this case, efficiency and predictive performance seem to be conflicting goals.

6.6 Summarizing Comments

One of the most popular paths to face multilabel classification lies in training a set of relatively simple classifiers, then joining their predictions to obtain the final output. Several MLC ensemble-based solutions have been described in the preceding sections of this chapter. These can be considered as more advanced proposals than the ensembles referenced in the previous chapters.

Fig. 6.6 Overview of
multilabel ensemble-based
methods

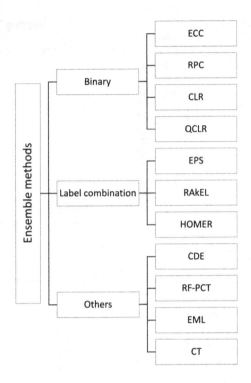

The diagram in Fig. 6.6 summarizes these ensemble-based methods, grouping them according to the type of underlying classifier they rely on, binary, multiclass, and other approaches. There are even ensembles made up by other ensembles, as is the case of EML which combines ECC, CLR, and RAkEL among other MLC methods.

In the second half of the chapter, four of the referenced ensemble methods have been experimentally tested. Two overall conclusions can be drawn from the analysis of results. The first one is that, on average, EPS is the ensemble that best deliver when it comes to predictive performance. The second is that the quickest ensemble, both to train the base classifiers and to make up the final prediction, is HOMER. However, these judgments should be taken with a grain of salt, since a limited number of methods and MLDs have been included in the experiments.

References

1. Tsoumakas, G., Dimou, A., Spyromitros, E., Mezaris, V., Kompatsiaris, I., Vlahavas, I.: Correlation-based pruning of stacked binary relevance models for multi-label learning. In: Proceedings of 1st International Workshop on Learning from Multi-Label Data, MLD'09, pp. 101–116 (2009)
2. Alvares-Cherman, E., Metz, J., Monard, M.C.: Incorporating label dependency into the binary relevance framework for multi-label classification. Expert Syst. Appl. **39**(2), 1647–1655 (2012)

3. Read, J., Pfahringer, B., Holmes, G., Frank, E.: Classifier chains for multi-label classification. Mach. Learn. **85**, 333–359 (2011)
4. Hüllermeier, E., Fürnkranz, J., Cheng, W., Brinker, K.: Label ranking by learning pairwise preferences. Artif. Intell. **172**(16), 1897–1916 (2008)
5. Fürnkranz, J., Hüllermeier, E., Loza Mencía, E., Brinker, K.: Multilabel classification via calibrated label ranking. Mach. Learn. **73**, 133–153 (2008)
6. Read, J., Pfahringer, B., Holmes, G.: Multi-label classification using ensembles of pruned sets. In: Proceedings of 8th IEEE International Conference on Data Mining, ICDM'08, pp. 995–1000. IEEE (2008)
7. Tsoumakas, G., Vlahavas, I.: Random k-Labelsets: An ensemble method for multilabel classification. In: Proceedings of 18th European Conference on Machine Learning, ECML'07, vol. 4701, pp. 406–417. Springer (2007)
8. Tsoumakas, G., Katakis, I., Vlahavas, I.: Effective and efficient multilabel classification in domains with large number of labels. In: Proceedings of ECML/PKDD Workshop on Mining Multidimensional Data, MMD'08, pp. 30–44 (2008)
9. Tenenboim-Chekina, L., Rokach, L., Shapira, B.: Identification of label dependencies for multi-label classification. In: Working Notes of the Second International Workshop on Learning from Multi-Label Data, MLD'10, pp. 53–60 (2010)
10. Kocev, D., Vens, C., Struyf, J., Džeroski, S.: Ensembles of multi-objective decision trees. In: Proceedings of 18th European Conference on Machine Learning, ECML'07, pp. 624–631. Springer (2007)
11. Tahir, M.A., Kittler, J., Bouridane, A.: Multilabel classification using heterogeneous ensemble of multi-label classifiers. Pattern Recogn. Lett. **33**(5), 513–523 (2012)
12. Read, J., Martino, L., Olmos, P.M., Luengo, D.: Scalable multi-output label prediction: from classifier chains to classifier trellises. Pattern Recogn. **48**(6), 2096–2109 (2015)
13. Madjarov, G., Gjorgjevikj, D., Džeroski, S.: Dual layer voting method for efficient multi-label classification. In: Proceedings of 5th Iberian Conference, IbPRIA'11, pp. 232–239. Springer (2011)
14. Mencía, E.L., Park, S., Fürnkranz, J.: Efficient voting prediction for pairwise multilabel classification. Neurocomputing **73**(7), 1164–1176 (2010)
15. Quevedo, J.R., Luaces, O., Bahamonde, A.: Multilabel classifiers with a probabilistic thresholding strategy. Pattern Recogn. **45**(2), 876–883 (2012)
16. Quinlan, J.R.: C4.5: Programs for machine learning (1993)

Chapter 7
Dimensionality Reduction

Abstract High dimensionality is a profoundly studied problem in machine learning. Usually, a high-dimensional input space defies most classification algorithms, tending to produce more complex and less effective models. Multilabel data are also affected by high dimensionality in the output space, since many datasets have hundreds or even thousands of labels. This chapter aims to explain how high dimensionality affects multilabel classification, as well as the methods proposed to deal with this obstacle. A general overview of the curse of dimensionality in the multilabel field is provided in Sect. 7.1. Section 7.2 introduces feature space reduction techniques, outlining several specific proposals and testing how applying feature selection impacts multilabel classifiers results. Then, a similar discussion but related to label space dimensionality is given in Sect. 7.3, also including some experimental results. Section 7.4 summarizes the chapter.

7.1 Overview

Classification is a data mining task which aims to recognize the category of new patterns. For doing so, the algorithms use some kind of model to compare the new sample input features with those of patterns already observed. The presence of redundant and irrelevant features impacts these algorithms, making them slower, due to the higher dimensionality, and usually also less precise, by cause of unrelated information contributed by irrelevant attributes. In the multilabel field, high dimensionality in the input space is something common, and it goes with high dimensionality also in the output space almost always.

The term *curse of dimensionality* was introduced in [1]. It denotes the problems derived from the presence of many variables (dimensions). As the number of dimensions increases, so does the volume of the solution space. As a consequence, data points in this volume tend to be more sparse as the dimensions grow, and distances between them tend to be less significant. Thus, to draw meaningful conclusions generally a larger collection of data points is needed. On the contrary, the accuracy of the estimation made by most algorithms will quickly degrade.

© Springer International Publishing Switzerland 2016   115
F. Herrera et al., *Multilabel Classification*,
DOI 10.1007/978-3-319-41111-8_7

In the data mining area, the curse of dimensionality term is used to make reference to problems where there are many input attributes, a trait that challenges most pattern recognition methods [14]. The raw data collected to generate new datasets usually includes superfluous features, with redundant information that can be derived from other attributes, as well as irrelevant data which are not meaningful to solve the problem at glance. Manually filter the attributes, in order to choose those which really provides relevant information, usually is very expensive. Therefore, automatic methods to accomplish this task have been designed over the years.

A problem of high dimensionality is always linked to the input feature space in traditional classification, since there are only one output attribute. However, the perspective is totally different while working with multilabeled data. It is usual to work with MLDs having hundreds or even thousands of labels, thus having to deal with a high-dimensional output space [3]. The difficulties produced by this configuration are similar to that of a high-dimensional input space, implying more time to build the classifiers and usually a degraded predictive performance. Though, the methods applied to reduce the input feature space are not appropriate for reducing the output label space.

This section briefly introduces the main obstacles derived from high-dimensional input and output spaces in multilabel datasets. In Sect. 7.2, several methods described in the literature to perform feature selection over multilabel data are portrayed, including some experiments. Section 7.3 is devoted to label reduction techniques, also comprising experimental results.

7.1.1 High-Dimensional Input Space

MLC techniques are used in fields such as text categorization and image labeling. These have a common characteristic, the large number of features extracted from the text, images, music, etc. Most MLDs have several hundreds, sometimes thousands, of attributes. High dimensionality in the input space is usually one or more orders of magnitude above with respect to traditional classification datasets. In general, the search space in MLC tasks is much bigger than in binary or multiclass ones.

Since the selection of the best subset of features [5] is mostly faced as a combinatorial problem, the larger is the number of attributes in the dataset the more time will be needed to find the ideal subset. In fact, evaluating all possible combinations can be unfeasible for many MLDs. Feature selection methods based on heuristics or statistical approaches can be applied in these cases.

An added barrier to reduce the input feature space in MLDs is derived from their own nature. Since there are multiple outputs to consider, instead of only one, determining the significance of every input attribute to each joint labelset is not an easy job. In many cases, the number of outputs can be larger than the number of inputs. This specific casuistic, along with the large set of input features in most MLDs, makes the problem of feature selection in MLC harder than usual.

7.1.2 High-Dimensional Output Space

Some of the problems of dealing with many labels are, at a certain extent, analogous to those faced in the input attribute space. A high-dimensional output space implies most complex models, for instance:

- All the algorithms based on the binarization techniques, such as the ones described in Chap. 4, would demand much more time to be built with large sets of labels. In this case, the complexity will increase linearly with respect to the number of labels with OVA approaches and quadratically with OVO.
- LP-based methods (see Chap. 4) would also face some obstacles. As the set of labels grows, the number of combinations increases exponentially. As a consequence, usually there are only a few data samples as representatives of each label combination. The sparseness tends to degrade the performance of many methods.
- As stated in Chap. 6, most ensemble MLC methods are based on BR and LP transformations, so the difficulties just described in the previous two bullets are also applicable in this case.
- Most classifiers have to build a model from the input features taking into account their relationship with the output labels. One of the few exceptions to this rule are the methods based on nearest neighbors. In general, the larger is the set of labels, the bigger and more complex would be that model.

The previous ones are problems that do not exist in traditional classification, so they open a new variety of preprocessing tasks that have to be specifically designed for multilabel data.

As a rule of thumb, reducing the output space, that is the number of labels, will contribute to decrease the time and memory needed to train most classifiers. In addition, those will be simpler and they could reduce overfitting, generalizing better. However, labels cannot be simply removed from the data. Therefore, the usual techniques followed to perform feature selection generally are not useful in the label space.

7.2 Feature Space Reduction

High dimensionality, as well as the term curse of dimensionality, is traditionally associated with the input attribute space. It is a broad and well-known problem which affects many machine learning tasks [18], including binary and multiclass classification, and obviously also multilabel classification. As a consequence, there are many proposed techniques to deal with this problem, all of them aimed to reduce the input space dimensionality. To do so, methods aimed to remove redundant and irrelevant features while preserving the most essential properties of the data have been designed. Many of them can be applied to multilabel data, straightly or following some kind of data transformation.

Besides computational complexity, and the aforementioned statement indicating that more features imply the need for more data instances, working in a high-dimensional input space also makes difficult data visualization tasks. Data with more than three dimensions can be hardly represented in a graphical way, being this an important tool in exploratory data analysis. Feature selection methods can be a way to simplify visualization of high-dimensional data.

7.2.1 Feature Engineering Approaches

Dimensionality reduction algorithms can be grouped into several categories, according to different criteria. Those briefly described below are among the most common ones:

- **Feature selection versus feature extraction**: Feature selection methods evaluate the relevance of attributes already present in the original data, selecting those which can provide more useful information for building a model. On the other hand, feature extraction [10] methods generate new attributes from the original ones. This kind of algorithms bore as a path to find the intrinsic dimensionality [27] of a dataset. Collectively, the process of selecting or building features is usually known as feature engineering.
- **Supervised versus unsupervised**: Unsupervised feature engineering aims to reduce dimensionality while preserving certain characteristics of the attributes, but without taking into account the class labels associated with each instance. The best-known unsupervised method is PCA (*Principal Component Analysis*) [13], whose goal is to reduce the number of attributes in the data but conserving their original variance. By contrast, supervised techniques take advantage of class labels and usually analyze correlations between each attribute and the class. CCA (*Canonical Correlation Information*) [11] and LDA (*Linear Discriminant Analysis*) [8] are such methods, determining the dependencies between inputs and outputs by means of cross-variance matrices. There are also semi-supervised feature selection methods proposed [31] in the literature.
- **Filter versus wrapper**: To reduce the number of attributes a method can rely exclusively on the original dataset, preprocessing it to produce a new version with less features as a result. Relying on some quality criteria, such as distance or divergence metrics, information measures, error probabilities, or consistency and distances between classes among others, which are not linked to any learning algorithm, there is not interaction between the biases of the feature selection and learning methods. This would be a filter method [4], whose main advantage is to be classifier independent. The formerly mentioned PCA and CCA algorithms are filter methods. Wrapper methods [15], on the other hand, are designed to optimize

the subset of features internally using a given classifier in the evaluation process. Therefore, this approach is classifier dependent, and it is able to infer attribute interactions with a specific classifier. A third path is embedded algorithms, which perform feature selection as part of the model training operation instead of as a separate stage. Decision trees and Random forest [2] work as embedded feature selection methods. Usually filter methods are more efficient, they need less time to do their work, while wrappers provide better performance.

- **Linear versus nonlinear**: Many of the most commonly used feature selection algorithms, such as the aforementioned PCA and LDA, are based on linear statistical methods such as linear regression. For instance, in PCA a linear transformation is applied over the original features, projecting them in a lower dimensionality space. Nonlinear methods [9] tend to provide more performance than linear ones, but at the expense of a higher computational complexity. Some of them are based on the kernel tricks and manifold embeddings.

Unsupervised dimensionality reduction methods rely on the analysis of the redundant information provided by the input features, for instance determining if some of them are linear combinations of others. Since they do not use the class label, these methods can be applied right out of the box to multilabel data. Supervised approaches, on the contrary, have to be adapted in some way to take into account the presence of multiple output labels. A general overview of feature selection and extraction methods can be found in [10, 18]. The following section is focused on algorithms specifically adapted to work with MLDs.

7.2.2 Multilabel Supervised Feature Selection

Traditional supervised feature selection and extraction, such as LDA or CCA, are designed to observe only one output class. The goal usually is to elucidate how much useful information carries each input feature to distinguish between the values of the class, obtaining a feature ranking from which a certain amount of them can be removed. So, there is a set of independent variables and only one dependent variable. The quality of each feature or subset of features is evaluated either by using an extensive collection of metrics in filter methods, such as Information gain and Relief [16], or according to the estimation provided by the underlying learning algorithm in wrapper methods. The scenario is a bit more complex in the multilabel field, since there are also a set of dependent variables, the labels, instead of only one.

It is possible to apply traditional feature selection methods to multilabel data, following some of the transformation approaches described in Chap. 4. For instance:

- **Binary relevance**: Firstly, the MLD is transformed into a set of binary datasets. Then, the contribution of the features to each individual label is evaluated. Lastly, the average contribution of the features to all labels is computed, and those who are above a certain threshold are chosen.

- **Label powerset**: After transforming the MLD into a multiclass dataset, by taking each label combination as class identifier, the feature selection algorithm is applied as usual. Where the BR-based approach does not consider potential label dependencies, the LP-based one implicitly includes this information in the process.

The use of these transformation-based methods, along with Information gain and Relief feature weighting measures, was reviewed in [21]. The conducted experimentation showed high variations in the number of features selected by each approach. Some transformation-based algorithms, such as the mutual information-based feature selection for multilabel classification introduced in [7], remove rare label combinations generated by the LP transformation and prune the space to explore for dependencies.

In addition to classic unsupervised methods and supervised ones implemented through transformation techniques, many others have been proposed to accomplish dimensionality reduction for MLDs. Some of them are briefly described below:

- **MI-based feature selection using interaction information**: Presented in [17], this method is based on mutual information between input features and output labels, but without performing any data transformation. The authors introduce a score metric able to measure interactions among inputs and outputs of any degree, analyze the bounds of this score, and use them in the algorithm to obtain the most relevant features.
- **Multilabel informed latent semantic indexing**: This proposal [28] follows the unsupervised approach, as the LSI (*Latent Semantic Indexing*) method it is based on, well known in the information retrieval field. However, the information related to labels is incorporated by applying to the label space exactly the same projection obtained for the feature space.
- **Memetic feature selection algorithm for multilabel classification**: Genetic algorithms (GAs) are well known for their ability to find near optimum solutions, based on a fitness function and evolution principles such as mutation and inheritance between one population and the following one. They also have some drawbacks, such as premature convergence and large computation times. The method proposed in [17] overcomes these obstacles resorting to memetic techniques, extracting the most relevant features with better efficiency and efficacy than GAs.
- **Multilabel dimensionality reduction via dependence maximization**: The authors of this method, introduced in [30], aimed to project the feature space into a lower dimensional space, like many other feature selection methods, while maximizing the dependence between the preserved features and the corresponding class labels. For doing so, they founded their method in the Hilbert–Schmidt independence criterion as metric to measure the dependencies among input attributes and output labels.
- **Hypergraph spectral learning**: Hypergraphs are graphs able to capture relations in high-dimensional spaces. In [22], a method to model the correlations among labels, based on an hypergraph, is proposed, along with a procedure to mild the computational complexity. The authors state that the result is equivalent

to a least-squares problem, and they use it to define an efficient algorithm to reduce the number of features in large-scale MLDs.

Beyond the methods just enumerated, several dimensionality reduction techniques and frameworks are described in [23], a recently published book fully devoted to this topic.

7.2.3 Experimentation

Most MLDs have large feature spaces, commonly much larger than binary and multi-class datasets. Feature engineering methods are applied to MLDs usually looking for two goals, a significant time reduction in the training and classification processes, and an improvement in the obtained predictions. In this section, a simple feature selection algorithm is going to be experimentally tested, aiming to verify the attainment of the aforementioned goals.

Five MLDs with a large number of input attributes have been selected. Their basic traits are those shown in Table 7.1. One of them, scene, has a few hundreds, while the others have more than a thousand features. All of them were preprocessed with a BR-Information gain feature selection algorithm. This procedure independently evaluates the quality of each feature with respect to each label, measuring the information gain. This is the same measure used by C4.5 to determine the best feature to divide a branch while inducing a tree. The scores for each feature across all labels is averaged, obtaining a feature ranking from which the best half is chosen. Therefore, the preprocessed version of the MLDs have half the input attributes than the original ones.

Both the original MLDs and preprocessed versions were given as input to three MLC methods, BR, LP, and CC. These have been already described in the previous chapters. The first one is a binary transformation, the second a label combination transformation, and the last an ensemble of binary classifiers. So classification results before and after preprocessing the MLDs have been obtained. Those results

Table 7.1 Basic traits of MLDs used in feature selection experimentation

Dataset	n	f	k	Card	Dens	TCS
bibtex	7 395	1 836	159	2.402	0.015	17.503
enron	1 702	1 001	53	3.378	0.064	20.541
genbase	662	1 186	27	1.252	0.046	13.840
medical	978	1 449	45	1.245	0.028	15.629
scene	2 407	294	6	1.074	0.179	10.183

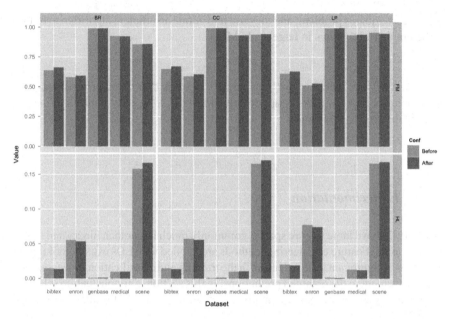

Fig. 7.1 Classifier performance before and after feature selection

are assessed with three performance metrics, *HL* (hamming loss), *FM* (F-Measure), and training time.

The plot in Fig. 7.1 summarizes the *HL* and *FM* predictive performance values. Remember that for *HL* lower is better, while for *FM* is the opposite. Grouped bars are used to show the results before and after preprocessing. At first sight, slight improvements for bibtex and enron seem to be obtained for both metrics. The differences for the other three MLDs, if any, are not very apparent in this graphic representation.

Raw values for each test configuration are provided in Table 7.2. Best results are highlighted in bold, as usual. According to the *FM* metric, the feature selection has improved classification results in all cases for four out of the five MLDs, being medical the exception with BR and CC, two binary-based classifiers, and scene with LP. The evaluation with *HL* is quite mixed, with improvements and diminishments in equal proportion.

Regarding the time spent training each model, the percentage of reduction with the preprocessed MLDs with respect to the original ones has been represented in Fig. 7.2. As can be seen, with the exception of genbase all the cases are around or above the 50 % threshold. This means that the preprocessed versions needed half of the time or less to train the classifier. As a consequence that the feature selection applied has been beneficial can be stated, since running time has been remarkably decreased while classification performance is maintained or even improved in many cases (Table 7.3).

Table 7.2 Classification results before and after label space reduction

		HL		FM	
Classifier	Dataset	Before	After	Before	After
BR	bibtex	0.0146	**0.0137**	0.6376	**0.6615**
	enron	0.0555	**0.0535**	0.5808	**0.5931**
	genbase	**0.0012**	0.0013	0.9908	0.9908
	medical	**0.0098**	0.0100	**0.9259**	0.9229
	scene	**0.1581**	0.1670	0.8568	**0.8589**
LP	bibtex	0.0204	**0.0192**	0.6113	**0.6294**
	enron	0.0776	**0.0745**	0.5138	**0.5291**
	genbase	0.0018	**0.0012**	0.9918	**0.9924**
	medical	0.0132	**0.0122**	0.9322	**0.9369**
	scene	**0.1658**	0.1679	**0.9547**	0.9453
CC	bibtex	0.0147	**0.0137**	0.6494	**0.6718**
	enron	0.0574	**0.0558**	0.5893	**0.6053**
	genbase	**0.0012**	0.0013	0.9908	0.9908
	medical	**0.0101**	0.0106	**0.9323**	0.9314
	scene	**0.1658**	0.1708	0.9378	**0.9419**

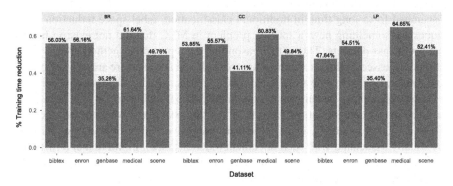

Fig. 7.2 Training time improvement after feature selection

Table 7.3 Training time reduction after feature selection

Dataset	BR (%)	LP (%)	CC (%)
bibtex	56.03	47.64	53.85
enron	56.16	54.51	55.57
genbase	35.28	35.40	41.11
medical	61.64	64.65	60.83
scene	49.76	52.41	49.84

7.3 Label Space Reduction

The total number of labels in an MLD, as well as the cardinality and other label-related characterization metrics introduced in Chap. 3, has a great influence in the classifiers behavior. Having more labels involves more individual classifiers for all the binarization-based methods, such as BR, CC, or RPC. A higher cardinality suggests the existence of more label combinations, impacting the LP-based methods, such as PS, RAkEL, or HOMER.

Intuitively, it seems that reducing the number of labels could help in decreasing the time needed to build the models and maybe improving their predictive performance. However, labels cannot be simply removed from an MLD, as the feature selection methods do with input attributes, since all labels have to be present in the prediction provided by the classifier. Thence, any label reduction method has to be able to recover or rebuild the labelsets to preserve their primitive structure.

7.3.1 Sparseness and Dependencies Among Labels

Unlike feature selection and extraction methods, which were already in place and researched for a long time in traditional classification, label space reduction techniques are something new, a specific problem in MLC that did not exist in binary and multiclass datasets. This is the reason that each proposed method, they will be further discussed, follows a different approach. Nonetheless, there are two key points to consider, label sparseness and label dependencies, whether separately or jointly.

Label sparseness is quite common in MLDs belonging to certain fields, for instance in text categorization. The classification vocabulary, which determines the words included as labels, tends to be very extensive, while only a few concepts are usually associated with each document (data instance). In multilabel terms, this means having MLDs with a large number of labels but a low cardinality; thus, the active labels are sparse in the label space. There are algorithms that can make the most of this sparseness, compressing the label space following a reversible approach.

One of the recurrent studied topic in the MLC field is label dependency. Some classification approaches, such as BR and similar methods, assume that labels are completely independent, and they perform quite well in many configurations. However, that labels which tend to appear together have some kind of relation or dependency [6] is presumed. It is usual to differentiate between two dependency models:

- **Unconditional dependence**: It is the dependence that presumably exists between certain labels, no matter what the input feature values are. Y_i and Y_j being two labels, they are unconditional dependent if $\mathcal{P}(Y_i|Y_j) \neq \mathcal{P}(Y_i)$. Unconditional dependence provides correlation information about labels in the MLD as a whole; thus, it can be globally exploited while designing an MLC method.
- **Conditional dependence**: This is the kind of dependence among labels according to the feature values of a specific data instance. Some algorithms, for instance

those based on label combinations such as LP, can take advantage of conditional dependence between labels, producing a model able to use this locally available information.

The authors of [29] highlight that dimensionality reduction taking advantage of label dependency information is one of the pending challenges in MLC. In the following section, some of the methods proposed in the literature for label space dimensionality reduction, most of them based on label space sparsity and label dependency principles, are depicted.

7.3.2 Proposals for Reducing Label Space Dimensionality

There are a handful of published methods whose aim is to reduce the dimensionality of the label space in multilabel classification. Each one of them follows a different approach to do so, with few exceptions. Some of the most remarkable are briefly described below:

- **Label subsets selection**: Some algorithms, such as RAkEL [25] and HOMER [24] (see Chap. 6), work by splitting the full labelsets into several subsets. Founded on the LP data transformation technique, these methods train a group of classifiers using a reduced set of labels for each one. RAkEL makes these subsets randomly, while HOMER relies on an hierarchical algorithm. Eventually, all the labels in the MLD are used to train one multiclass classifier, since the label space is not effectively reduced but only divided into groups.

- **Pruning of infrequent labelsets**: The PS [19] and EPS [20] methods (see Chap. 4), also based on the basic LP transformation approach, detect rare labelsets and prune them. This way the main problem of LP-based algorithms, which is the combinatorial explosion as the number of labels gets larger, is relieved at some extent. As RAkEL and HOMER, PS and EPS also rely on multiclass classifiers to produce their predictions. The infrequent labelsets pruning avoids taking into account rare label combinations, but seldom reduce the number of labels.

- **Kernel Dependency Estimation (KDE)**: KDE is a generic technique to find dependencies among a set of inputs and a set of outputs. In [26] is proposed as a way to reduce the label space for MLC tasks. This method applies a principal component analysis to the original label space, obtaining a set of uncorrelated projections. After that, the number of outputs is reduced keeping only the most significant ones, those with larger eigenvalues. The proposal is completed with a procedure to rebuild the original labelset once the reduced prediction has been obtained, based on a method to find the preimage of the projected version.

- **Compressed sensing (CS)**: The authors of [12] proposed a method to use CS, a compression technique, in order to reduce the dimensionality of the label space for multilabel classification. The original binary high-dimensional label space is projected into a real lower dimensional compressed space. Since the intermediate values are not binary but real, the authors resort to a regression algorithm to generate the predictions, instead of a classifier. Eventually, the outputs of regression models have to be decompressed to produce the predicted labelsets. This technique has a premise that the label space presents a considerable sparsity level is assumed.

- **Compressed labeling on distilled labelsets** (**CL**): This is a variant of the CS method described above and introduced in [32]. First, it extracts from the original MLD the most frequent sets of labels. These are the so-called distilled labelsets. Then, the label space is projected into a lower dimensional space, but using the signs of the random projections to preserve the binary nature of the original one. This way the resulting problem can be faced with a classification algorithm, instead of regression. The partial predictions produced by the classifier are finally complemented by using the correlation information stored in the distilled labelsets.
- **Label inference for addressing high dimensionality in the label space**: The proposal in [3] is called LI-MLC, and it is based on taking advantage of label dependency information. To obtain this information, an association rule mining algorithm is used. Taking the labels as items and the instances as transactions, a set of association rules is generated. The antecedent of each rule indicates which labels must appear in the prediction to infer that the label in the consequent also should be present. Keeping only the most confident rules, those labels whose presence can be deduced from others are removed. The resulting problem is still a multilabel classification task, but with a reduced label space. Therefore, any MLC algorithm can be used to generate the predictions. These will be complemented in the final step by evaluating the association rules, inferring the missing ones from the presence of others.

As can be seen, some of these methods, such as LI-MLC and CL, gather label dependency information to evaluate which of the labels can be temporarily removed. Others, as CS and CL, assume a certain level of sparseness in the label space, on the contrary the compression scheme they rely on cannot be applied.

7.3.3 Experimentation

In Chap. 3, while studying the characteristics of several dozens of MLDs, that many of them have hundreds or even thousands of labels was stated. In some cases, indeed the set of labels can be larger than the set of input features. The implications of working with a high number of labels have been already mentioned. Therefore, by reducing this number an algorithm aims to improve classification results and usually also running time.

This final section intends to experimentally test one of the aforementioned label space reduction methods, assessing how it impacts the behavior of some MLC classifiers. The chosen method is LI-MLC, the algorithm which banks on association rules as representation model of label dependency information. Since the goal is to analyze a label reduction algorithm, five MLDs with a relatively large set of labels have been used. These are `cal500`, `corel16k`, `corel5k`, `delicious` and `imdb`.

The last one only has 28 labels, while the other four have several hundreds of them, with delicious reaching almost a thousand labels. The basic traits of these MLDs are provided in Table 7.4. The full set of characteristics can be found in the tables in Chap. 3.

The experimental configuration is the same used before for testing the feature selection method. Thus, classification results before and after applying LI-MLC with three different classifiers, BR, LP, and CC, and two evaluation metrics, *HL* and *FM*, have been analyzed. Furthermore, the running time of each version has been also obtained, including the time spent in extracting the association rules.

Classifier performance has been represented in Fig. 7.3 as a set of bar plots. From the *FM* values observation (top half of the figure), that a general improvement in classification results has been achieved can be deducted. The only exception seems

Table 7.4 Basic traits of MLDs used in label space reduction experimentation

Dataset	n	f	k	Card	Dens	TCS
cal500	502	68	174	26.044	0.150	15.597
corel16k	13 766	500	153	2.859	0.019	19.722
corel5k	5 000	499	374	3.522	0.009	20.200
delicious	16 105	500	983	19.017	0.019	22.773
imdb	120 919	1 001	28	2.000	0.071	18.653

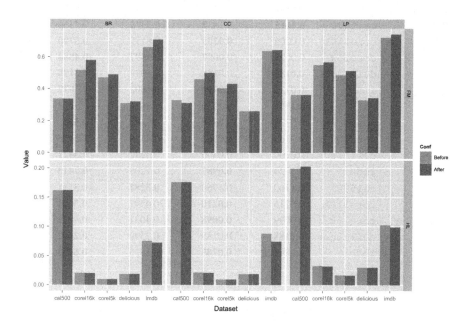

Fig. 7.3 Classifier performance before and after label reduction

to be the `cal500` MLD. The lower half corresponds to *HL* measures; thus, lower values are better. Again, `cal500` shows the worst performance, while the results for the other four MLDs suggest the same behavior showed by *FM*.

The exact values for each classifier and MLD, with and without using the LI-MLC label selection method, are the shown in Table 7.5. As can be seen, the values highlighted in bold for the *FM* metric all belong to the configuration which includes LI-MLC. As was anticipated, `cal500` is the only exception and their results sign a slight worsening. The behavior when assessing with the *HL* metric are very similar. Only `delicious` is negatively affected in some cases, but the differences are almost negligible.

With regard to the running times, the relative gain (or loss) is presented in Fig. 7.4. Negative values indicate that the use of LI-MLC plus the classifier took longer than the classifier alone. This happened with `imbd`, the MLD which has more samples. The association rule mining algorithm needed more time to analyze the instances than the reduction obtained by removing some labels. On the other hand, with `corel16k` the number of removed labels compensates the time mining the rules; thus, the global running time is significantly improved. In general, BR and CC are more influenced by the reduction of labels than LP (Table 7.6).

Table 7.5 Classification results before and after label space reduction

Classifier	Dataset	HL		FM	
		Before	After	Before	After
BR	cal500	**0.1615**	0.1620	**0.3375**	0.3364
	corel16k	0.0206	**0.0203**	0.5155	**0.5790**
	corel5k	0.0098	**0.0097**	0.4697	**0.4884**
	delicious	**0.0186**	0.0188	0.3097	**0.3198**
	imdb	0.0753	**0.0724**	0.6602	**0.7077**
LP	cal500	**0.1996**	0.2030	**0.3626**	0.3627
	corel16k	0.0329	**0.0318**	0.5496	**0.5663**
	corel5k	0.0168	**0.0162**	0.4859	**0.5125**
	delicious	0.0300	0.0300	0.3290	**0.3421**
	imdb	0.1024	**0.0987**	0.7209	**0.7413**
CC	cal500	0.1760	0.1760	**0.3287**	0.3111
	corel16k	0.0216	**0.0210**	0.4592	**0.5001**
	corel5k	0.0099	**0.0098**	0.4031	**0.4305**
	delicious	**0.0188**	0.0189	0.2599	**0.2602**
	imdb	0.0878	**0.0745**	0.6375	**0.6433**

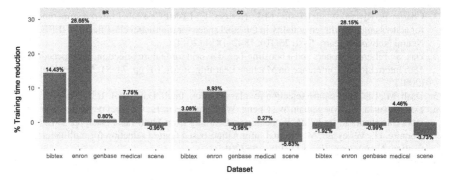

Fig. 7.4 Training time improvement after label space reduction

Table 7.6 Training time reduction after label selection

Dataset	BR (%)	LP (%)	CC (%)
cal500	14.43	3.08	−1.92
corel16k	28.65	8.93	28.15
corel5k	0.80	−0.98	−0.99
delicious	7.75	0.27	4.46
imdb	−0.95	−5.63	−3.73

7.4 Summarizing Comments

This chapter has been devoted to dimensionality reduction in multilabel data. In addition to the usual high-dimensional feature space, already known in traditional classification, MLC algorithms also have to deal with a high-dimensional label space. Sometimes the amount of labels in an MLD can even outnumber the whole set of input features. How these two characteristics influence different MLC solutions has been analyzed, and several feature selection and label reduction techniques have been described.

The conducted experimentation has tested the impact which a feature selection method and a label reduction one have in a selection of three classifiers and five datasets. In general, it seems that the applying of feature selection had a bigger impact in running times, while performing label selection tends to improve more classification results. Although here these two techniques have been used separately, they could be combined.

References

1. Bellman, R.: Dynamic Programming. P (Rand Corporation). Princeton University Press (1957)
2. Breiman, L.: Random forests. Mach. Learn. **45**(1), 5–32 (2001)

3. Charte, F., Rivera, A., del Jesus, M.J., Herrera, F.: LI-MLC: a label inference methodology for addressing high dimensionality in the label space for multilabel classification. IEEE Trans. Neural Networks Learn. Syst. **25**(10), 1842–1854 (2014)

4. Das, S.: Filters, wrappers and a boosting-based hybrid for feature selection. In: Proceedings of 18th International Conference on Machine Learning, ICML'01, pp. 74–81. Morgan Kaufmann (2001)

5. Dash, M., Liu, H.: Feature selection for classification. Intell. Data Anal. **1**(3), 131–156 (1997)

6. Dembszynski, K., Waegeman, W., Cheng, W., Hüllermeier, E.: On label dependence in multi-label classification. In: ICML Workshop on Learning from Multi-label Data, pp. 5–12 (2010)

7. Doquire, G., Verleysen, M.: Mutual information-based feature selection for multilabel classification. Neurocomputing **122**, 148–155 (2013)

8. Fisher, R.A.: The statistical utilization of multiple measurements. Ann. Eugenics **8**(4), 376–386 (1938)

9. Guyon, I., Bitter, H.M., Ahmed, Z., Brown, M., Heller, J.: Multivariate non-linear feature selection with kernel multiplicative updates and Gram-Schmidt relief. In: Proceedings of International Joint Workshop on Soft Computing for Internet and Bioinformatics, BISC Flint-CIBI'03, pp. 1–11 (2003)

10. Guyon, I., Gunn, S., Nikravesh, M., Zadeh, L.A. (eds.): Feature Extraction: Foundations and Applications. Springer (2008)

11. Hotelling, H.: Relations between two sets of variates. In: Breakthroughs in Statistics, pp. 162–190. Springer (1992)

12. Hsu, D., Kakade, S., Langford, J., Zhang, T.: Multi-label prediction via compressed sensing. In: Proceedings of 22th Annual Conference on Advances in Neural Information Processing Systems, NIPS'09, vol. 22, pp. 772–780 (2009)

13. Jolliffe, I.: Principal Component Analysis. Springer Series in Statistics, vol. 1. Springer, Berlin (1986)

14. Kira, K., Rendell, L.A.: The feature selection problem: traditional methods and a new algorithm. In: Proceedings of 10th National Conference on Artificial Intelligence, AAAI'92, pp. 129–134. AAAI Press (1992)

15. Kohavi, R., John, G.H.: Wrappers for feature subset selection. Artif. Intell. **97**(1), 273–324 (1997)

16. Kononenko, I.: Estimating attributes: analysis and extensions of RELIEF. In: Machine Learning: ECML-94, pp. 171–182 (1994)

17. Lee, J.S., Kim, D.W.: Mutual information-based multi-label feature selection using interaction information. Expert Syst. Appl. **42**, 2013–2025 (2015)

18. Liu, H., Motoda, H.: Feature Selection for Knowledge Discovery and Data Mining, vol. 454. Springer Science & Business Media (2012)

19. Read, J.: A pruned problem transformation method for multi-label classification. In: Proceedings of New Zealand Computer Science Research Student Conference, NZCSRS'08, pp. 143–150 (2008)

20. Read, J., Pfahringer, B., Holmes, G.: Multi-label classification using ensembles of pruned sets. In: Proceedings of 8th IEEE International Conference on Data Mining, ICDM'08, pp. 995–1000. IEEE (2008)

21. Spolaor, N., Cherman, E.A., Monard, M.C., Lee, H.D.: A comparison of multi-label feature selection methods using the problem transformation approach. Electron. Notes Theor. Comput. Sci. **292**, 135–151 (2013)

22. Sun, L., Ji, S., Ye, J.: Hypergraph spectral learning for multi-label classification. In: Proceedings of 14th ACM SIGKDD International Conference on Knowledge Discovery and Data Mining, pp. 668–676. ACM (2008)

23. Sun, L., Ji, S., Ye, J.: Multi-Label Dimensionality Reduction. CRC Press (2013)

24. Tsoumakas, G., Katakis, I., Vlahavas, I.: Effective and efficient multilabel classification in domains with large number of labels. In: Proceedings of ECML/PKDD Workshop on Mining Multidimensional Data, MMD'08, pp. 30–44 (2008)

25. Tsoumakas, G., Vlahavas, I.: Random k-Labelsets: an ensemble method for multilabel classi-fication. In: Proceedings of 18th European Conference on Machine Learning, ECML'07, vol. 4701, pp. 406–417. Springer (2007)
26. Weston, J., Chapelle, O., Elisseeff, A., Schölkopf, B., Vapnik, V.: Kernel dependency estima-tion. In: Proceedings of 16th Annual Conference on Advances in Neural Information Processing Systems, NIPS'02, vol. 15, pp. 873–880 (2002)
27. Wyse, N., Dubes, R., Jain, A.K.: A critical evaluation of intrinsic dimensionality algorithms. Pattern Recogn. Pract. 415–425 (1980)
28. Yu, K., Yu, S., Tresp, V.: Multi-label informed latent semantic indexing. In: Proceedings of 28th Annual International ACM SIGIR Conference on Research and Development in Information Retrieval, pp. 258–265. ACM (2005)
29. Zhang, M., Zhou, Z.: A review on multi-label learning algorithms. IEEE Trans. Knowl. Data Eng. **26**(8), 1819–1837 (2014)
30. Zhang, Y., Zhou, Z.H.: Multilabel dimensionality reduction via dependence maximization. ACM Trans. Knowl. Discovery Data (TKDD) **4**(3), 14 (2010)
31. Zhao, Z., Liu, H.: Semi-supervised feature selection via spectral analysis. In: Proceedings of 7th SIAM International Conference on Data Mining, SDM'07, pp. 641–646 (2007)
32. Zhou, T., Tao, D., Wu, X.: Compressed labeling on distilled labelsets for multi-label learning. Mach. Learn. **88**(1–2), 69–126 (2012)

Chapter 8
Imbalance in Multilabel Datasets

Abstract The frequency of class labels in many datasets is not even. On the contrary, that a certain class appears in a large portion of the data samples while other is scarcely represented is something quite usual. This casuistic produces a problem generically labeled as class imbalance. Due to these differences between class distributions, a specific need arises, imbalanced learning. This chapter beings introducing the mentioned task in Sect. 8.1. Then, the specific aspects of imbalance in the multilabel area are discussed in Sect. 8.2. Section 8.3 explains how imbalance in MLC has been faced, enumerating a considerable set of proposals. Some of them are experimentally evaluated in Sect. 8.4. Lastly, Sect. 8.5 summarizes the contents.

8.1 Introduction

Learning from imbalanced data is a challenge for many classification algorithms. Since most classifiers are designed to minimize a certain global error measurement, when they have to deal with imbalanced data, they tend to benefit the most frequent class. Miss-classification of rare classes does not have a great impact in the global performance assessment conducted by most evaluation metrics. However, depending on the scenario, the main interest of the task could be on correctly label these rare patterns, instead of the most common ones.

Imbalanced learning is a well-studied problem in the binary and multiclass scenarios [10, 13, 16, 19, 22]. The imbalance level in binary datasets is computed as the ratio between the most frequent or majority class and the less frequent one or minority class. It is the so-called Imbalance Ratio (*IR*), later adapted to work with multiclass datasets.

The imbalanced learning task has been faced mostly following one of three approaches:

- **Data resampling**: Resampling techniques are usually implemented as a pre-processing step, thus producing a new dataset from the original one. To balance the class distribution, it is possible to remove instances associated with the majority class or to generate new samples linked to the minority class [18]. Resampling methods are mostly classifier independent, so they can be seen as a general

© Springer International Publishing Switzerland 2016   133
F. Herrera et al., *Multilabel Classification*,
DOI 10.1007/978-3-319-41111-8_8

solution to this problem. Nonetheless, there are also some resampling proposals
for specific classifiers.

- **Algorithm adaptation**: This approach is classifier dependent. Its goal is to modify
  existent classification algorithms to take into account the imbalanced nature of the
  data to be processed. The usual procedure is based on reinforcing the learning of
  the minority class, biasing the classifier to recognize it.
- **Cost-sensitive learning**: Cost-sensitive classification is an approach which combines the two previous techniques. The data are preprocessed to balance the class
  distribution, while the learning algorithm is adapted to benefit correct classification of samples associated with the minority class. To do so weights are associated
  with the instances, and usually these weights are proportional to the size of each
  class.

From these three ideas, many others have been derived, such as the combination
of data resampling and the use of ensembles of classifiers [11] as a more robust model
with certain tolerance to class imbalance.

Overall, imbalance learning is a well-known and deeply studied task in binary
classification, further extended to also cover the multiclass scenario. Imbalance in
multilabeled data increases the complexity of the problem and potential solutions,
since there are several class labels per instance. In the following, the specificities
of imbalanced MLDs, related problems, and proposed methods to tackle them are
described.

> Most of the existent methods only consider the presence of one majority class
> and one minority class. This way, undersampling methods only remove samples
> from one class, and analogously oversampling methods generate new instances
> associated with one class.

8.2   Imbalanced MLD Specificities

The number of labels in an MLD can go from a few dozens to several thousands.
Only a handful of them have less than ten labels. Despite the fact that most MLDs
have a large set of labels, the average number of active labels per instance (their
cardinality) seldom is above 5. Some exceptions are $cal500$ ($Card = 26.044$) and
$delicious$ ($Card = 19.017$). With such a large set of labels and low $Card$, that
some labels would be underrepresented while others would be much more frequent
can be deducted. As a general rule, the more labels there are in an MLD, the higher
would be the likelihood of having imbalance problems.

Another important fact, easily deducible from the own MLDs nature, is that there
is not a single majority label and a single minority one, but several of them in each
group. This have different implications, affecting the way the imbalance level of an

MLD can be measured or the behavior of resampling and classification methods, as will be further detailed in the following sections of this chapter.

The way in which multilabel classification is faced can make worse the imbalance problem. Transformation techniques such as BR sometimes produce extreme imbalance levels. The binary dataset corresponding to a minority class will only have a few instances representing it, while all the others will belong to the opposite class. On the other hand, the LP transformation has to deal with rare label combinations, those in which the scarce minority labels appear, on their own or jointly with some of the majority ones. All the BR- and LP-based methods will face similar problems.

8.2.1 How to Measure the Imbalance Level

The metrics related to imbalance measurement for MLDs were provided in Chap. 3 (see Sect. 3.3.2). Since there are multiple labels, this trait cannot be easily reduced to a single value. For that reason, a triplet of metrics was proposed in the study conducted in [5]:

- *IRLbl*: It is independently computed for each label. The value of this metric will be 1 for the most frequent label and higher for all others. The larger is the *IRLbl* the less frequent is the assessed label in the MLD.
- *MeanIR*: By averaging the *IRLbl* for all labels in a MLD, its *MeanIR* is obtained. This value typically will be above 1. The higher is the *MeanIR*, the more imbalanced labels there are in the MLD.
- *CVIR*: The *MeanIR* is intended to give a measurement on the amount of imbalanced labels in the MLD, but it is also influenced by extreme values. A few very high-imbalanced labels can produce a high *MeanIR*, the same that a lot of less imbalanced labels. The *CVIR* is an indicator of the situation being assessed. Large *CVIR* values would denote high variances in *IRLbl*.

Besides the use of specific characterization metrics, such as the ones just described, one of the best approaches to analyze label imbalance in MLDs is to visually explore the data. In Fig. 8.1, the relative frequencies for the ten most frequent labels (left side) and the ten least frequent ones (right side) in a dozen MLDs have been plotted.[1] As can be observed, the difference between frequent and rare labels is huge. Even among the most frequent labels, there are significant disparities, with one or two labels having much more presence than the others. This pattern is common to many MLDs. Therefore, the imbalance problem is almost intrinsically linked to multilabel data.

---

[1] The frequency (Y-axis) scale is individually adjusted to show better the relevance of labels in each MLD, instead of being common to all plots.

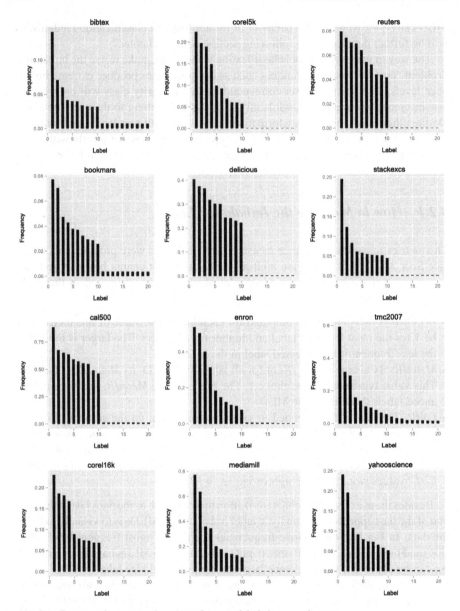

Fig. 8.1 Ten most frequent and ten least frequent labels in some datasets

8.2.2   Concurrence Among Imbalanced Labels

Looking at the graphical representations on Fig. 8.1, as well as to the imbalance levels reported in the tables on Chap. 3, it seems legitimate to think that applying resampling methods, as in traditional classification, the labels distribution on the MLDs could be balanced. However, MLDs have a specific characteristic which is not present on traditional datasets. As we are already aware, each data sample is associated with several outputs, and some of them can be minority labels while others are majority ones.

Due to this peculiarity, entitled as concurrence among imbalanced labels in [3], resampling methods could be not as effective as they should. In the same paper, a specific metric to assess this casuistic, named *SCUMBLE*, is proposed. It was defined in Sect. 3.3.3. As was demonstrated in this study, MLDs with large *SCUMBLE* values, that is with a high concurrence between minority and majority labels, usually do not benefit from resampling techniques as much as MLDs without this problem.

Visualizing the concurrence among imbalanced labels is not easy, since most MLDs have too many labels to show them at once along with their interactions.

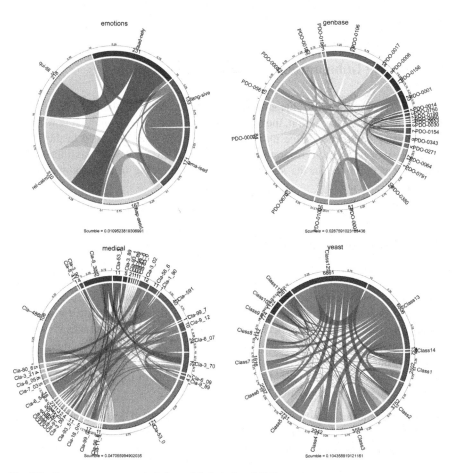

Fig. 8.2 Concurrence among imbalance labels in four MLDs

Nonetheless, it is possible to limit the number of labels to show, choosing those with higher dependencies, producing plots such as the ones shown[2] in Fig. 8.2.

Each arc in the external circumference represents a label. The arc's amplitude is proportional to the frequency of the label, so small arcs are associated with minority labels, and analogously large arcs indicate majority labels. The width of the bands connecting arcs denote the number of samples in which each label pair appears together.

> Multilabel imbalance-aware methods able to take into account label concurrence could potentially produce better results than those that do not consider this information. A further section details such a method developed by the authors, called REMEDIAL.

8.3   Facing Imbalanced Multilabel Classification

On the basis of the specific characteristics associated with imbalanced MLDs, highlighted in the previous section, the design of algorithms capable of dealing with this problem is a challenge. Three main approaches have been followed in the literature, classifier adaptation, resampling methods, and ensembles of classifiers. Most of them are portrayed in the subsections below according to the aforementioned categorization scheme.

8.3.1   Classifier Adaptation

One way to face the imbalance problem consists in adapting the classifier to take this aspect into consideration, for instance assigning weights to each label depending on its frequency. Obviously, it is a solution tightly attached to the adjusted algorithm. Although it is not a general application approach, what can be seen as a disadvantage, the adaptation can strengthen the best point of a good classifier, something that a preprocessing method cannot do.

Some of the multilabel classifiers adapted to deal with imbalanced MLDs proposed in late years are the following:

- **Min–max modular with SVM (M³-SVM):** This method, proposed in [8], relies on a Min–Max Modular network [17] to divide the original multilabel classification problem into a set of simpler tasks. Several strategies are used to guarantee that the

---

[2]These plots were generated by the mldr R package, described in the following chapter.

imbalance level in these smaller tasks is lower than in the original MLD, following random, clustering, and PCA approaches. The simpler tasks are always binary classification jobs, using SVM as base classifier. Therefore, the proposal can be seen as a combination of data transformation and method adaptation techniques.

- **Enrichment process for neural networks**: The proposal made in [25] is an adaptation of the training process for neural networks. This task is divided into three phases. The first one uses a clustering method to group similar instances and gets a balanced representation to initialize the neural network. In the second stage, the network is iteratively trained, as usual, while data samples are added and removed from the training set, according to their prevalence. The final phase checks if the enrichment process has reached the stop condition or it has to be repeated. This way, the overall balance of the neural network used as classifier is improved.

- **Imbalanced multimodal multilabel learning (IMMML)**: In [14], the authors face an extremely imbalanced multilabel task, specifically the prediction of subcellular localization of human proteins. Their algorithm is based on a Gaussian process model, combined with latent functions on the feature space and covariance matrices to obtain correlations among labels. The imbalance problem is tackled giving each label a weighting coefficient linked to the likelihood of labels on each sample. Therefore, it is very specific solution to a definite problem, hardly applicable in a different context.

- **Imbalanced multiinstance multilabel radial basis function neural networks (IMIMLRBF)**: It was introduced in [15] as an extension to the MIMLRBF algorithm [26], a multiinstance and multilabel classification algorithm based on radial basis neural networks. The adaptation consists in two key points. Firstly, the number of units in the hidden layer, which in MIMLRBF is constant, is computed according to the number of samples of each label. Secondly, the weights associated with the links between the hidden and output layers are adjusted, biasing them depending on the label frequencies.

8.3.2 Resampling Techniques

The resampling approach is based on removing samples which belong to the majority label, adding samples associated with the minority label, or both actions at once. The way the instances to be removed are selected, and the technique used to produce new instances, usually follows one of two possible ways. The first one is the random approach, whereas the second one is known as heuristic approach. The former randomly chooses the data samples to delete, imposing the restriction that they have to belong to a certain label. Analogously, new samples are produced by randomly picking and cloning instances associated with a specific label. The latter path can be based on disparate heuristics to search the proper instances, as well as to generate new ones.

Therefore, resampling methods can be grouped depending on the way they try to balance the label frequency, removing or adding samples, and the strategy to do so, randomly or heuristically. There are quite a few proposals based on resampling techniques, among them:

- **Undersampling for imbalanced training sets in text categorization domains**: The proposal made in [9] combines the data transformation approach, producing a set of binary classifiers, with undersampling techniques, removing instances linked to the majority label to balance the distribution in each binary dataset. In addition, a decision tree is used to get the most relevant features for each label. kNN is used as underlying binary classifier, and different k values were tested in the conducted experimentation.

Algorithm 1 LP-RUS algorithm's pseudo-code.

   Inputs: <Dataset> D, <Percentage> P
   Outputs: Preprocessed dataset
1: $samplesToDelete \leftarrow |D|/100 * P$ $\triangleright$ P% size reduction
2: $\triangleright$ Group samples according to their labelsets
3: **for** $i = 1 \rightarrow |labelsets|$ **do**
4:    $labelSetBag_i \leftarrow$ samplesWithLabelset(i)
5: **end for**
6: $\triangleright$ Calculate the average number of samples per labelset
7: $meanSize \leftarrow 1/|labelsets| * \sum\limits_{i=1}^{|labelsets|} |labelSetBag_i|$
8: $\triangleright$ Obtain majority labels bags
9: **for each** $labelSetBag_i$ **in** $labelSetBag$ **do**
10:    **if** $|labelSetBag_i| > meanSize$ **then**
11:       $majBag_i \leftarrow labelSetBag_i$
12:    **end if**
13: **end for**
14: $meanRed \leftarrow samplesToDelete/|majBag|$
15: $majBag \leftarrow$ SortFromSmallestToLargest($majBag$)
16: $\triangleright$ Calculate # of instances to delete and remove them
17: **for each** $majBag_i$ **in** $majBag$ **do**
18:    $rBag_i \leftarrow \min(|majBag_i| - meanSize, meanRed)$
19:    $remainder \leftarrow meanRed - rBag_i$
20:    distributeAmongBags$_{j>i}$($remainder$)
21:    **for** $n = 1 \rightarrow rBag_i$ **do**
22:       $x \leftarrow$ random($1, |majBag_i|$)
23:       deleteSample($majBag_i, x$)
24:    **end for**
25: **end for**

- **LP-based resampling (LP-ROS/LP-RUS)**: In [2], two resampling methods, named LP-ROS (*Label Powerset Random Oversampling*) and LP-RUS (*Label Powerset Random Undersampling*), are presented. As their names suggest, they do not evaluate the frequency of individual labels, but of full labelsets. LP-RUS removes instances from the most frequent labelsets, whereas LP-ROS clones samples associated with the least frequent ones. The pseudo-code for the LP-RUS algorithm is shown in Algorithm 1. As can be seen, the algorithm takes as input the percentage of samples to remove from the MLD. After computing the average number of samples sharing each labelset, a set of majority bags are produced. The number of instances to delete is distributed among these majority bags, randomly picking the data samples to remove. The LP-ROS algorithm works in a very similar fashion, but obtaining bags with minority labelsets and adding to them clones of samples randomly retrieved from them. These are simple techniques, and they consider the presence of several majority and minority combinations, instead of only one as most resampling methods assume.

Algorithm 2 ML-ROS algorithm's pseudo-code.

Inputs: \<Dataset\> D, \<Percentage\> P
Outputs: Preprocessed dataset
1: $samplesToClone \leftarrow |D|/100 * P$ ▷ P% size increment
2: $L \leftarrow$ labelsInDataset(D)   ▷ Obtain the full set of labels
3: $MeanIR \leftarrow$ calculateMeanIR(D, L)
4: **for each** $label$ **in** L **do**   ▷ Bags of minority labels samples
5:     $IRLbl_{label} \leftarrow$ calculateIRperLabel($D, label$)
6:     **if** $IRLbl_{label} > MeanIR$ **then**
7:        $minBag_{i++} \leftarrow Bag_{label}$
8:     **end if**
9: **end for**
10: **while** $samplesToClone > 0$ **do**   ▷ Instances cloning loop
11:     ▷ Clone a random sample from each minority bag
12:     **for each** $minBag_i$ **in** $minBag$ **do**
13:        $x \leftarrow$ random($1, |minBag_i|$)
14:        cloneSample($minBag_i, x$)
15:        **if** $IRLbl_{minBag_i} <= MeanIR$ **then**
16:           $minBag \rightarrow minBag_i$ ▷ Exclude from cloning
17:        **end if**
18:        $--samplesToClone$
19:     **end for**
20: **end while**

- **Random resampling by label (ML-ROS/ML-RUS)**: As in the previous study, two resampling methods are also introduced in [5], one for oversampling and another one for undersampling. Both evaluate the individual imbalance level per label, deleting instances linked to the majority labels (ML-RUS) or cloning those associated with the minority ones (ML-ROS). The imbalance level is assessed by means of the *IRLbl* and *MeanIR* metrics defined in [2]. The removing/cloning process is iterative, and it reassess the imbalance levels in each iteration aiming to achieve the best balance for all labels. ML-ROS increases the number of instances in a given percentage, by cloning those with minority labels, while ML-RUS does the opposite by removing majority labels. The pseudo-code for ML-ROS is provided in Algorithm 2. Once the number of clones to produce is computed, the *IRLbl* and *MeanIR* are used to get a bag with the instances in which each minority label appears. The clones will be generated from these bags, following the iterative approach aforementioned. A new sample is created from each minority bag, reassessing their condition of minority bags in each cycle. This way, the best possible balance for each group is set as goal. The ML-RUS algorithm behavior is quite similar, but it gets bags with majority labels and iteratively removes samples from them.

- **A case study with the SMOTE algorithm**: The authors of the study published in [12] stated the imbalance problem in MLC, and proposed to face it using the original SMOTE (*Synthetic Minority Over-sampling Technique*) algorithm [18]. SMOTE was designed to produce synthetic instances of the minority class for binary datasets. In [12], three ways to feed SMOTE with multilabel data are tested, all of them considering one minority label only. The first path is similar to BR, giving to SMOTE all the instances in which the minority label appears to obtain synthetic samples from them and their neighbors. The second approach is quite limited, since only considers instances having the minority label alone, without any other labels. The third way, which probed to be the most effective, grouped the minority label instances according to the combinations of labels in which it appeared.

Algorithm 3 MLeNN algorithm pseudo-code.

Inputs: <Dataset> D, <Threshold> HT, <NumNeighbors> NN
Outputs: Preprocessed dataset

```
1: for each sample in D do
2:     for each label in getLabelset(D) do
3:         if IRLbl(label) > MeanIR then
4:             Jump to next sample                    ▷ Preserve instance with minority labels
5:         end if
6:     end for
7:     numDifferences ← 0
8:     for each neighbor in nearestNeighbors(sample, NN) do
9:         if adjustedHammingDist(sample, neighbor) > HT then
10:            numDifferences ← numDifferences+1
11:        end if
12:    end for
13:    if numDifferences≥NN/2 then
14:        markForRemoving(sample)
15:    end if
16: end for
17: deleteAllMarkedSamples(D)
```

- **Multilabel edited nearest neighbor (MLeNN)**: MLeNN is an heuristic undersampling algorithm. The method is proposed in [4], and it is build upon the well-known ENN (*Edited Nearest-Neighbor*) rule [21], foundation of a simple data cleaning procedure. It compares the class of each instance against the one of its NNs, usually its three NNs. Those samples whose class differs from the class of two or more NNs are marked for removing. The algorithm, presented in [4] and whose pseudo-code is shown in Algorithm 3, adapts the ENN rule to the MLC field introducing two key ideas, a principle to chose the samples acting as candidates to be removed and a comparison operator to determine when the labelsets of two instances are considered to be different. Only the instances which do not contain any minority label are used as candidates, instead of all the samples as in the original ENN implementation. Regarding how the classes of these instances are compared, a metric based on the Hamming distance among labelsets, but only taking into account active labels, is defined.

Algorithm 4 MLSMOTE algorithm's pseudo-code.

Inputs:
   D ▷ Dataset to oversample
   k ▷ Number of nearest neighbors

1: $L \leftarrow$ labelsInDataset(D)   ▷ Full set of labels
2: $MeanIR \leftarrow$ calculateMeanIR(D, L)
3: **for each** $label$ **in** L **do**
4:   $IRLbl_{label} \leftarrow$ calculateIRperLabel($D, label$)
5:   **if** $IRLbl_{label} > MeanIR$ **then**
6:    ▷ Bags of minority labels samples
7:    $minBag \leftarrow$ getAllInstancesOfLabel($label$)
8:    **for each** $sample$ **in** $minBag$ **do**
9:     $distances \leftarrow$ calcDistance($sample, minBag$)
10:     sortSmallerToLargest($distances$)
11:     ▷ Neighbor set selection
12:     $neighbors \leftarrow$ getHeadItems($distances, k$)
13:     $refNeigh \leftarrow$ getRandNeighbor($neighbors$)
14:     ▷ Feature set and labelset generation
15:     $synthSmpl \leftarrow$ newSample($sample$,
16:         $refNeigh, neighbors$)
17:     $D = D + synthSmpl$
18:    **end for**
19:   **end if**
20: **end for**

- **Multilabel SMOTE (MLSMOTE):** This is another MLC oversampling method based on the SMOTE algorithm. However MLSMOTE, the proposal introduced in [6], is a full adaptation of the original algorithm toward the use of MLDs, instead of a procedure to use the unchanged SMOTE method with multilabel data as proposed in [12]. MLSMOTE considers several minority labels, instead of only one, taking the samples in which these labels appear as seeds to generate new data instances. To do so, firstly their nearest neighbors are found and the input features are obtained by interpolation techniques. Thus, the new instances are synthetic rather than mere clones of existing samples. Three approaches are tested to produce the synthetic labelsets associated with the new samples. Two of them rely on set operations among the labelsets of the NNs, computing the union or the intersection of active labels. The third one, eventually the one that produced better results, generates a ranking of labels in the NNs, keeping those present on half or more of the neighbors. As can be seen in Algorithm 4, corresponding to the main body of the MLSMOTE algorithm, the method relies on the *IRLbl* and *MeanIR* measurements to extract a collection of minority bags, each one corresponding to a label. Then, the k-nearest neighbors are retrieved. One of them will be used to reference instance to produce the synthetic features, while the labels on all of them (see Algorithm 5) serve to generate the synthetic labelset.

Algorithm 5 Function: Generation of new synthetic instances.

21: **function** NEWSAMPLE(*sample, refNeigh, neighbors*)
22:    *synthSmpl* ← **new** Sample   ▷ New empty instance
23:    ▷ Feature set assignment
24:    **for each** *feat* **in** *synthSmpl* **do**
25:       **if** typeOf(*feat*) is numeric **then**
26:          *diff* ← *refNeigh.feat* - *sample.feat*
27:          *offset* ← *diff* * randInInterval(0,1)
28:          *value* ← *sample.feat* + *offset*
29:       **else**
30:          *value* ← mostFreqVal(*neighbors,feat*)
31:       **end if**
32:       *syntSmpl.feat* ← *value*
33:    **end for**
34:    ▷ Label set assignment
35:    *lblCounts* ← counts(sample.labels)
36:    *lblCounts* + ← counts(neighbors.labels)
37:    *labels* ← *lblCounts* > *(k+1)* / 2
38:    *synthSmpl.labels* ← *labels*
39:    **return** *synthSmpl*
40: **end function**

- **Resampling by decoupling highly imbalanced labels (REMEDIAL)**: None of the above resampling methods consider the concurrence among imbalanced labels, the problem previously described in Sect. 8.2.2. This is the differential factor of REMEDIAL, the method presented in [1] and whose pseudo-code is shown in Algorithm 6. It is an algorithm specifically designed to work with MLDs having a high *SCUMBLE*, the metric used to assess the concurrence level. It works both as an oversampling method and as an editing procedure. Firstly, the instances with high *SCUMBLE* values, those in which minority and majority labels appear together, are located. Then, for each sample in the previous set a new sample is produced by preserving the original features, but containing only minority labels. Lastly, the original sample is edited by removing these same minority labels. This way, the samples which can make harder the learning process are decoupled. As the authors highlight in [1], this method can be used as a previous step to apply other resampling techniques.

Algorithm 6 REMEDIAL algorithm.

1: **function** REMEDIAL(MLD D, Labels L)
2:   $IRLbl_l \leftarrow$ calculateIRLbl(l in L)   ▷ Calculate imbalance levels
3:   $IRMean \leftarrow \overline{IRLbl}$
4:   $SCUMBLEIns_i \leftarrow$ calculateSCUMBLE(D_i in D)   ▷ Calculate SCUMBLE
5:   $SCUMBLE \leftarrow \overline{SCUMBLEIns}$
6:   **for each** *instance i* **in** D **do**
7:     **if** $SCUMBLEIns_i > SCUMBLE$ **then**
8:       $D'_i \leftarrow D_i$ ▷ Clone the affected instance
9:       $D_i[labels_{IRLbl<=IRMean}] \leftarrow 0$ ▷ Maintain minority labels
10:      $D'_i[labels_{IRLbl>IRMean}] \leftarrow 0$ ▷ Maintain majority labels
11:      $D \leftarrow D + D'_i$
12:     **end if**
13:   **end for**
14: **end function**

The main advantage of these methods is that they are classifier independent. They are used as a preprocessing step, even it is possible to combine them, and they do not require a specific multilabel classifier to be used. Therefore, the preprocessed MLDs can be later given as input to any of the MLC algorithms described in previous chapters.

8.3.3 The Ensemble Approach

Ensemble-based techniques are quite common in the MLC field. A significant number of proposals have been already published, as was reported in Chap. 6 devoted to multilabel ensembles. ECC, EPS, RAkEL, and HOMER are among the most popular MLC ensembles, an approach that also has been applied to solve the imbalance problem.

Theoretically, each classifier in an ensemble could introduce a bias toward a different set of labels, easing and making more effective the imbalanced learning task. The following two proposals are headed in this direction:

- **Inverse random undersampling (BR-IRUS)**: The method proposed in [24] is built upon an ensemble of binary classifiers. Several of them are trained for each label, using a subset of the original data with each one. This subset of the instances includes all samples in which the minority label is present, as well as a small portion of the remainder samples. This way, each individual classifier faces a balanced classification task. Joining the predictions given by the classifiers associated with a label, a more defined boundary around the minority label space is generated. The name of the proposal, BR-IRUS, highlights the fact of using the binary relevance transformation.

- **Ensemble of multilabel classifiers (EML)**: Developed by the same authors of the previous work, in [23] the construction of an heterogeneous ensemble of multilabel classifiers to tackle the imbalance problem is introduced. The ensemble is made up of five classifiers. All of them are trained with the same data, but using different algorithms. The underlying MLC classifiers chosen by the authors are RAkEL, ECC, CLR, MLkNN, and IBLR. Several methods for joining the individual predictions are tested, along with different thresholding and weighting schemes width adjustments made through cross-validation.

Usually, the major drawback of ensembles is their computational complexity, since a set with several classifiers has to be trained and their predictions have to be combined. This obstacle is more substantial in the case of EML, as the base classifiers are ensembles by themselves. In addition, these solutions are not classifier independent, being closer to the classifier adaptation approach than to resampling techniques.

8.4 Multilabel Imbalanced Learning in Practice

In the previous sections, most of the published methods aimed to tackle multilabel imbalanced learning have been portrayed. The goal in this section is to experimentally test some of them. Five methods, belonging to different techniques, have been chosen, specifically:

- **Random resampling**: Two algorithms based on random resampling techniques have been applied, ML-RUS and ML-ROS. The former performs undersampling, by removing samples associated with majority labels randomly picking them, while the latter does the opposite, producing clones of instances linked to minority labels.
- **Heuristic resampling**: This group of approaches is also represented by two methods, MLeNN and MLSMOTE. The first one removes instances with majority labels following the ENN rule. The second produces synthetic instances associated with minority labels, generating both features and labelsets from the information in the neighborhood.
- **Ensembles**: The EML (ensemble-based method), just described in the previous section, is also included in the test bed. Unlike the previous four algorithms, EML is not a preprocessing technique but a full classifier by itself, able to face imbalance by combining predictions coming from several classifiers with different biases.

These five methods[3] have been tested using the experimental configuration explained in the following section. Obtained results are presented and discussed in Sect. 8.4.2.

---

[3]The implementations of these methods can be found in the links section provided in this book repository [7], along with dataset partitions.

Table 8.1 Basic traits of MLDs used in the experimentation

Dataset	n	f	k	Card	Dens	MeanIR
bibtex	7 395	1 836	159	2.402	0.015	12.498
cal500	502	68	174	26.044	0.150	20.578
corel5k	5 000	499	374	3.522	0.009	189.568
medical	978	1 449	45	1.245	0.028	89.501
tmc2007	28 596	49 060	22	2.158	0.098	15.158

8.4.1 Experimental Configuration

Four out of the five imbalance methods to be tested are preprocessing procedures. Therefore, once they have done their work, producing the rebalanced MLD, the data has to be given to an MLC classifier in order to obtain comparable classification results. A basic BR transformation has been used for this duty, with the C4.5 [20] tree induction algorithm as underlying binary classifier.

Five MLDs with disparate imbalance levels, bibtex, cal500, corel5k, medical and tmc2007, have been included in the experimentation. Their basic traits, including the *MeanIR*, are provided in Table 8.1. Each MLD was partitioned with a 2 × 5 fold cross-validation scheme, as usual. Training partitions were preprocessed with ML-RUS, ML-ROS, MLeNN, and MLSMOTE.

Thus, five versions of each one were used, one without resampling and four more preprocessed by each method. The original version, without resampling, was given as input to the BR classifier to obtain a base evaluation. It was also used with EML, which did not need an independent classifier. The preprocessed versions also served as input to the same BR + C4.5 MLC, with exactly the same configuration parameters.

In Chap. 3, the metrics designed to assess MLC algorithms performance were introduced. Many of them, such as *Hamming Loss*, *Accuracy*, *Precision*, *Recall*, and *F-measure*, have been used in the experiments of previous chapters. To study the behavior of classifiers when working with imbalanced data, as it is done here, it is usual to rely on label-based metrics, instead of sample-based ones. In this case, *F-measure* following the macro- and microaveraging strategies are the metrics obtained to assess the results. *MacroFM* (Macro-F-measure) assigns the same weight to all labels, while *MicroFM* (Micro-F-measure) is heavily influenced by the frequencies of each label. Therefore, the former is usually used to assess the performance with respect to minority labels, and the latter to obtain a more general view of the classifier's behavior.

8.4.2 Classification Results

Classification results assessed with *MacroFM* are shown in Fig. 8.3. Each group of bars corresponds to an MLD, with each bar depicting the performance of a method.

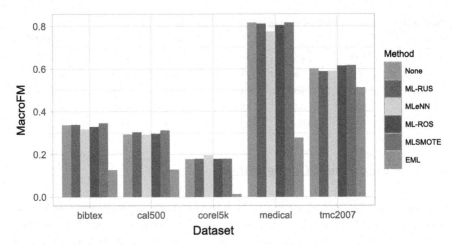

Fig. 8.3 Classification results assessed with the Macro-FMeasure metric

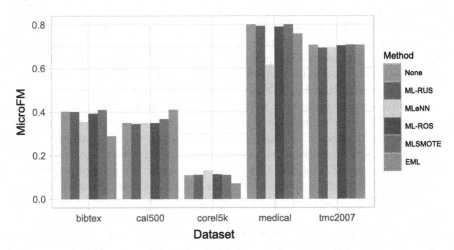

Fig. 8.4 Classification results assessed with the Micro-FMeasure metric

The left-most bar shows the value for base results, those obtained without any special imbalance treatment.

Analogously, Fig. 8.4 presents the results evaluated with the *MicroFM* metric. The structure of the plot is exactly the same. To analyze these data, it would be interesting to know in which cases an imbalance treatment has achieved some improvement over the base results. Another important fact is which one of the applied methods works better.

As can be observed in the two previous plots, undersampling methods seem to behave worse than the oversampling ones. The exception is MLeNN with the `corel5k` MLD, which achieves the best results with the two evaluation metrics.

Table 8.2 Results assessed with *MacroFM* (higher is better)

Dataset	Base	ML-RUS	MLeNN	ML-ROS	MLSMOTE	EML
bibtex	0.3368	*0.3383*	0.3170	0.3288	**0.3457**	0.1265
cal500	0.2933	*0.3029*	0.2918	*0.2966*	**0.3124**	0.1291
corel5k	0.1774	*0.1792*	**0.1966**	*0.1784*	*0.1790*	0.0133
medical	0.8165	0.8117	0.7750	0.8046	**0.8165**	0.2770
tmc2007	0.6015	0.5878	0.5903	*0.6138*	**0.6165**	0.5122

Table 8.3 Results assessed with *MicroFM* (higher is better)

Dataset	Base	ML-RUS	MLeNN	ML-ROS	MLSMOTE	EML
bibtex	0.4021	0.4007	0.3533	0.3927	**0.4097**	0.2888
cal500	0.3488	0.3447	0.3478	0.3478	*0.3663*	**0.4106**
corel5k	0.1096	*0.1109*	**0.1315**	*0.1135*	*0.1103*	0.0712
medical	0.8006	0.7935	0.6149	0.7902	**0.8006**	0.7581
tmc2007	0.7063	0.6934	0.6947	0.7038	**0.7071**	*0.7065*

EML, the ensemble-based solution, does not produce good *MacroFM* results, although with *MicroFM* the performance seems to be slightly better, specifically with the cal500 MLD. Regarding the oversampling methods, MLSMOTE appears as the best performed almost always. In fact, this method accomplishes the best results in many cases.

The *MacroFM* and *MicroFM* raw values are provided in Tables 8.2 and 8.3, respectively. Values highlighted in italics denote an amelioration with respect to results without imbalance treatment. Best values across all methods are emphasized in bold, as usual.

From these values observation it can be stated that EML seldom reaches the performance of the BR + C4.5 base classifier, although it achieves the best *MicroFM* result with the cal500 MLD. In comparison, MLSMOTE improves base results always for both metrics and manages to get the best performance in seven out of ten configurations. ML-RUS and ML-ROS produce some improvements, as well as a few losses. Lastly, MLeNN seems to work well with the corel5k MLD, but its behavior with the other four datasets is not as good.

Overall, it seems that advanced preprocessing techniques, such as the MLSMOTE algorithm, are able to improve MLC results while dealing with imbalanced MLDs.

8.5 Summarizing Comments

Class imbalance is a very usual obstacle while learning a classification model. In this chapter, how label imbalance is present in most MLDs, and some of the specificities, in this field such as label concurrence among imbalanced labels, have been introduced. Several metrics aimed to assess these traits have been explained, and some specialized data visualizations have been provided.

Solutions to deal with imbalanced multilabel data can be grouped into a few categories, including preprocessing methods, algorithm adaptation, and ensembles. A handful of proposals from each category have been described, and some of them have been experimentally tested. According to the results obtained, the resampling techniques deliver certain improvements while maintaining the benefit of being classifier independent.

References

1. Charte, F., Rivera, A., del Jesus, M.J., Herrera, F.: Resampling multilabel datasets by decoupling highly imbalanced labels. In: Proceedings of 10th International Conference on Hybrid Artificial Intelligent Systems, HAIS'15, vol. 9121, pp. 489–501. Springer (2015)
2. Charte, F., Rivera, A.J., del Jesus, M.J., Herrera, F.: A first approach to deal with imbalance in multi-label datasets. In: Proceedings of 8th International Conference on Hybrid Artificial Intelligent Systems, HAIS'13, vol. 8073, pp. 150–160. Springer (2013)
3. Charte, F., Rivera, A.J., del Jesus, M.J., Herrera, F.: Concurrence among imbalanced labels and its influence on multilabel resampling algorithms. In: Proceedings of 9th International Conference on Hybrid Artificial Intelligent Systems, HAIS'14, vol. 8480. Springer (2014)
4. Charte, F., Rivera, A.J., del Jesus, M.J., Herrera, F.: MLeNN: a first approach to heuristic multilabel undersampling. In: Proceedings of 15th International Conference on Intelligent Data Engineering and Automated Learning, IDEAL'14, vol. 8669, pp. 1–9. Springer (2014)
5. Charte, F., Rivera, A.J., del Jesus, M.J., Herrera, F.: Addressing imbalance in multilabel classification: measures and random resampling algorithms. Neurocomputing **163**, 3–16 (2015)
6. Charte, F., Rivera, A.J., del Jesus, M.J., Herrera, F.: MLSMOTE: approaching imbalanced multilabel learning through synthetic instance generation. Knowl. Based Syst. **89**, 385–397 (2015)
7. Charte, F., Rivera, A.J., del Jesus, M.J., Herrera, F.: Multilabel Classification. Problem analysis, metrics and techniques book repository. https://github.com/fcharte/SM-MLC
8. Chen, K., Lu, B., Kwok, J.: Efficient classification of multi-label and imbalanced data using min-max modular classifiers. In: Proceedings of IEEE International Joint Conference on Neural Networks, IJCNN'06, pp. 1770–1775 (2006)
9. Dendamrongvit, S., Kubat, M.: Undersampling approach for imbalanced training sets and induction from multi-label text-categorization domains. In: New Frontiers in Applied Data Mining. LNCS, vol. 5669, pp. 40–52. Springer (2010)
10. Fernández, A., López, V., Galar, M., del Jesus, M.J., Herrera, F.: Analysing the classification of imbalanced data-sets with multiple classes: binarization techniques and ad-hoc approaches. Knowl. Based Syst. **42**, 97–110 (2013)
11. Galar, M., Fernández, A., Barrenechea, E., Bustince, H., Herrera, F.: An overview of ensemble methods for binary classifiers in multi-class problems: experimental study on one-vs-one and one-vs-all schemes. pattern Recogn. **44**(8), 1761–1776 (2011)

12. Giraldo-Forero, A.F., Jaramillo-Garzón, J.A., Ruiz-Muñoz, J.F., Castellanos-Domínguez, C.G.: Managing imbalanced data sets in multi-label problems: a case study with the SMOTE algorithm. In: Proceedings of 18th Iberoamerican Congress on Progress in Pattern Recognition, Image Analysis, Computer Vision, and Applications, CIARP'13, vol. 8258, pp. 334–342. Springer (2013)

13. He, H., Garcia, E.A.: Learning from imbalanced data. IEEE Trans. Knowl. Data Eng. **21**(9), 1263–1284 (2009)

14. He, J., Gu, H., Liu, W.: Imbalanced multi-modal multi-label learning for subcellular localization prediction of human proteins with both single and multiple sites. PloS One **7**(6), 7155 (2012)

15. Li, C., Shi, G.: Improvement of learning algorithm for the multi-instance multi-label RBF neural networks trained with imbalanced samples. J. Inf. Sci. Eng. **29**(4), 765–776 (2013)

16. López, V., Fernández, A., García, S., Palade, V., Herrera, F.: An insight into classification with imbalanced data: empirical results and current trends on using data intrinsic characteristics. Inf. Sci. **250**, 113–141 (2013)

17. Lu, B., Ito, M.: Task decomposition and module combination based on class relations: a modular neural network for pattern classification. IEEE Trans. Neural Netw. **10**(5), 1244–1256 (1999)

18. Nitesh, V.C., Kevin, W.B., Lawrence, O.H., Kegelmeyer, W.P.: SMOTE: synthetic minority over-sampling technique. J. Artif. Intell. Res. **16**, 321–357 (2002)

19. Prati, R.C., Batista, G.E., Silva, D.F.: Class imbalance revisited: a new experimental setup to assess the performance of treatment methods. Knowl. Inf. Syst. **45**(1), 247–270 (2015)

20. Quinlan, J.R.: C4.5: Programs for Machine Learning (1993)

21. Sheskin, D.J.: Handbook of Parametric and Nonparametric Statistical Procedures. Chapman & Hall (2003)

22. Sun, Y., Wong, A.K.C., Kamel, M.S.: Classification of imbalanced data: a review. Int. J. Pattern Recogn. Artif. Intell. **23**(4), 687–719 (2009)

23. Tahir, M.A., Kittler, J., Bouridane, A.: Multilabel classification using heterogeneous ensemble of multi-label classifiers. Pattern Recogn. Lett. **33**(5), 513–523 (2012)

24. Tahir, M.A., Kittler, J., Yan, F.: Inverse random under sampling for class imbalance problem and its application to multi-label classification. Pattern Recogn. **45**(10), 3738–3750 (2012)

25. Tepvorachai, G., Papachristou, C.: Multi-label imbalanced data enrichment process in neural net classifier training. In: Proceedings of IEEE International Joint Conference on Neural Networks, IJCNN'08, pp. 1301–1307. IEEE (2008)

26. Zhang, M., Wang, Z.: MIMLRBF: RBF neural networks for multi-instance multi-label learning. Neurocomputing **72**(16), 3951–3956 (2009)

Chapter 9
Multilabel Software

Abstract Multilabel classification and other learning from multilabeled data tasks
are relatively recent, with barely a decade of history behind them. When compared
against binary and multiclass learning, the range of available datasets, frameworks,
and other software tools is significantly more scarce. The goal of this last chapter
is to provide the reader with the proper insight to take advantage of these software
tools. A brief overview of them is offered in Sect. 9.1. Section 9.2 discusses the
different multilabel file formats, enumerates the data repositories the MLDs can be
downloaded from, and describes how to automate some tasks with the mldr.datasets
R package. How to perform exploratory data analysis of MLDs is the main topic
of Sect. 9.3. Then, the process to conduct experiments with multilabel data using
different tools is outlined in Sect. 9.4.

9.1 Overview

Despite the software shortage aforementioned above, currently there are some multilabel data repositories, as well as two frameworks for algorithm developers and at
least one exploratory data tool. By using them, tasks such as downloading, citing and
partitioning datasets, multilabel data exploration, and conducting experiments with
existent MLC algorithms will be at your fingertips.

The present chapter has been structured into three main sections. The first one
describes the tools needed to work with multilabel data. This includes details about
MLDs file formats, data repositories MLDs can be obtained from, and how most
of these tasks can be accomplished by means of a specific software tool, the
mldr.datasets R package.

How to perform exploratory analysis of multilabel data is the topic the second
section is dedicated to. To do so, two specific programs are depicted, the R mldr
package and the Java MEKA framework. Many of the plots in this book have been
produced by the former, a tool which also provides methods to filter and transform
MLDs.

The concern of the third and final section is how to conduct multilabel experiments, by means of MEKA, MULAN, and a specific Java utility developed by the

© Springer International Publishing Switzerland 2016

F. Herrera et al., *Multilabel Classification*,

DOI 10.1007/978-3-319-41111-8_9

authors. Following these guidelines, and using the data partitions provided in this book repository, the reader should be able to reproduce the experiments described in previous chapters.

9.2 Working with Multilabel Data

The design of any algorithm aimed to deal with multilabeled data, whether its goal is to induce a classification model or to apply some kind of preprocessing, has a key requirement, it will have to be tested against some MLDs. Therefore, whatever is the researching goal, the first step will usually be obtaining enough multilabel data. These MLDs will have to be partitioned, and commonly some exploratory analysis would have to be conducted on them. In addition, they have to be properly documented into the research projects they are used, including the correct citing information.

Fortunately, nowadays there are several data repositories and software tools to fulfill these needs. This first section provides a brief description of such resources, along with useful references to obtain them.

9.2.1 Multilabel Data File Formats

One of the first issues that any multilabel researcher or practitioner (the user henceforward) has to face is the disparate set of MLDs file formats. Unlike traditional datasets, MLDs have more than one output attribute, so that the last feature is the class label cannot be assumed. How to communicate which ones of the features are labels is the main origin of the several file formats, because each developer came with a different approach to solve this obstacle.

Most MLDs are written using one of two base file formats, CSV (*Comma-Separated Values*) or ARFF[1] (*Attribute-Relation File Format*). Both are text file formats, but the latter includes a header with data descriptors followed by the data itself, whereas the former usually only provides the data and the header, if it is present, only brings field names. CSV files cannot contain label information, so the knowledge of how many labels there are, where are they located, or what are their names will depend on an external resource. By contrast, an ARFF file can include this information into the header.

---

[1]An ARFF file is usually divided into three sections. The first one contains the name of the dataset after de `@relation` tag, the second one provides information about the attributes with `@attribute` tags, and the third one, whose beginning is marked with the `@data` tag, contains the actual data. It is the file format used by the popular WEKA data mining tool.

MLDs can be downloaded from repositories (see the next section) such as MULAN [1], MEKA [2], LibSVM [3], KEEL [4], and RUMDR [5], each one using a different file format. The differences among multilabel file formats can be grouped according to the following criteria:

- **CSV versus ARFF**: ARFF is the most usual base file format for MLDs. The datasets available at MULAN, MEKA, and KEEL are ARFF files. On the contrary, LibSVM chose to use the simpler CSV file format.
- **Label information**: In order to use an MLD, knowing how many labels there are or which are the names of the attributes acting as labels is essential. MULAN datasets provide the label names in a separate XML file. KEEL datasets include in the ARFF header the set of output attributes. MEKA datasets indicate in the header, along with the name of the relation, the number of labels.
- **Label location**: Although multilabel formats providing label names could locate the attributes acting as labels at any position in the MLD, they usually put them at the end, after all the input attributes. This is the case for MULAN and KEEL. On the other hand, MEKA and LibSVM always arrange the labels at the beginning. Knowing the number of labels, the location allows to get the proper attribute names without needing to include them in the ARFF header or providing an XML file.
- **Sparse versus non-sparse**: There are MLDs that have thousands of input attributes plus thousands of labels. Therefore, each data row (instance) consists of a long sequence of values. Many of them could be zero, since labels can only take two values and the same is also applicable to many input attributes. In these cases, the MLD will be a large array, with thousands of columns and maybe rows, with zeroes in most of its values. To avoid storage and memory wasting, these MLDs are usually stored as sparse data. The rows in a sparse MLD are composed of comma-separated pairs of values. In each pair, the first value indicates the index of the attribute, while the second provides the actual value. In non-sparse MLDs, each row will contain the same number of columns, having the values for each attribute.

When some experiment is going to be conducted using a set of MLDs, the user has to choose between converting all of them to the proper file format, suitable for the tool to be used later to conduct the experiment, or being limited to only use those MLDs which are already available in this file format.

9.2.2 Multilabel Data Repositories

When it comes to multilabel data gathering, there are several alternatives to choose from. Custom MLDs can be produced for specific fields where multilabel data are still not available. Alternately, existing MLDs produced by someone else can be obtained from several repositories, as long as they suit the faced task needs. The option to generate these MLDs synthetically, by means of some software tools, is another potential choice. This section will look into the second approach.

Multilabel data repositories provide a convenient way to obtain MLDs that other researchers have built and used in their studies. It is an approach that allows to compare different strategies against the same data. Nonetheless, only full datasets are available some times. Few of these repositories also provide citation information. Therefore, the user usually has to download the MLD, partition it, and search for the proper bibliographic entry.

The following are among the best-known multilabel data repositories. For each one, the file format of the MLDs is also indicated:

- **MULAN**: MULAN [1] is a reference multilabel software tool (it will be further described), and its associated repository [6] is probably the most used resource by researchers in this field. The MLDs are provided in the ARFF format. The labels are usually located at the end of the attribute list, and each MLD is linked to an XML file containing the label names and their hierarchical relationship if it exists. Currently, this repository holds 27 MLDs,[2] some of them with prebuilt partitions.
- **MEKA**: MEKA is a multilabel tool based on WEKA. As MULAN, it brings reference implementations for several methods, as will be shown later. The MEKA repository [2] supplies 15 MLDs. Some of them are the same found in MULAN, but using the MEKA file format. This is also ARFF-based, but the labels always appear at the beginning of the attribute list. There are no separate XML file with label names, but the number of labels in the MLD is indicated in the ARFF header, as a parameter of the relation name.
- **LibSVM**: LibSVM [3] is a popular software library for SVMs. There are many classification algorithms, including some multilabel ones, build upon LibSVM. The associated data repository [7] includes 8 MLDs. In this case, the file format is CSV-based instead of ARFF-based, but the attribute values are given according to the sparse representation previously described. The labels are always put at the beginning of each instance. There is no associated XML file nor any header indicating the number of labels or their names.
- **KEEL**: Unlike MULAN and MEKA, KEEL [4] is a general-purpose data mining application, similar to WEKA. This software tool has an extensive data repository with different kinds of datasets, including 16 MLDs [8]. The file format is ARFF-based, indicating in the attribute list which features act as labels.
- **RUMDR**: The *R Universal Multilabel Dataset Repository* [5] is associated with an R package named mldr.datasets [9] (it will be portrayed in the following subsection). Currently, this is the most extensive multilabel data repository, providing more than 60 MLDs. These can be directly downloaded from the repository in R file format; thus, they are designed to be loaded from R. The functions provided by the package allow to export them from this native format to several ones, including MULAN, MEKA, LibSVM, and CSV.

---

[2]The number of MLDs provided by each repository has been checked as of April 2016.

- **Extreme Classification Repository**: This repository [10] only provides 9 MLDs, all of them sharing a specific characteristic: They have a huge list of input features, output labels, or both. There are MLDs with more than one million attributes, aimed to test solutions for extreme multilabel classification. The file format is a combination of CSV for label indexes, always at the beginning, and sparse representation for input attributes, with a one-line header indicating the number of instances, features, and labels.

These data repositories offer an immediate solution to the user which needs some MLDs, as long as the file format is appropriate and a tool to partition the data are at hand. However, this is not always the case. Depending on the tool being used to conduct the experiments, the MLDs may have to be transformed to other file format and properly partitioned. Some of these needs can be addressed by means of the software package described below.

9.2.3 The mldr.datasets Package

When it comes to data exploration, analysis, and mining, R [11] is a very popular tool/language due to its extensive package list. One of these packages, named mldr.datasets [9], is specifically designed to aid the user in the tasks of obtaining, partitioning, converting, and exporting multilabel datasets. mldr.datasets is tied to the aforementioned RUMDR repository.

The mldr.datasets is available at CRAN (*Comprehensive R Archive Network*), the distributed network providing most R packages. Therefore, it can be downloaded and installed from any up-to-date R version by simply issuing the console the `install.packages("mldr.datasets")` command. Once installed, the package has to be loaded into memory with the usual `library(mldr.datasets)` command. This will bring to the R workspace ten medium-sized MLDs, along with the functions needed to access many more and to manipulate them. The preloaded MLDs are those stored in the `data` folder of the RUMDR repository.

In the following, how to use the mldr.datasets to accomplish some basic tasks over MLDs is explained, assuming the package is already installed and loaded into memory.

9.2.3.1 Loading Available MLDs

After loading the package, the user can know which MLDs are available using the usual `data()` function, passing the name of the package as parameter. These are the MLDs brought to memory by loading the package, but there are many more available on the RUMDR repository. A list of these is returned by the `mldrs()` function, as shown in Fig. 9.1.

```
Data sets in package 'mldr.datasets':

birds                        Dataset with sounds produced by birds and the species
                             they belong to
cal500                       Dataset with music data along with labels for emotions,
                             instruments, genres, etc.
emotions                     Dataset with features extracted from music tracks and
                             the emotions they produce
flags                        Dataset with features correspoinding to world flags
genbase
langlog
medical
ng20

slashdot
stackex_chess
```

List of additional datasets available at the mldr.datasets repository   —   □   ✕

Archivo

	X	Name	Description
1	1	bibtex	Dataset with BibTeX entries
2	2	bookmarks	Dataset with data from web bookmarks and their ca>
3	3	corel16k001	Datasets with data from the Corel image collectio>
4	4	corel16k002	Datasets with data from the Corel image collectio>
5	5	corel16k003	Datasets with data from the Corel image collectio>
6	6	corel16k004	Datasets with data from the Corel image collectio>
7	7	corel16k005	Datasets with data from the Corel image collectio>
8	8	corel16k006	Datasets with data from the Corel image collectio>
9	9	corel16k007	Datasets with data from the Corel image collectio>
10	10	corel16k008	Datasets with data from the Corel image collectio>
11	11	corel16k009	Datasets with data from the Corel image collectio>
12	12	corel16k010	Datasets with data from the Corel image collectio>
13	13	corel5k	Dataset with data from the Corel image collection
14	14	delicious	Dataset generated from the del.icio.us site bookm>
15	15	enron	Dataset with email messages and the folders where>
16	16	eurlexdc_test	List with 10 folds of the test data from the EUR>
17	17	eurlexdc_tra	List with 10 folds of the train data from the EU>
18	18	eurlexev_test	List with 10 folds of the test data from the EUR>
19	19	eurlexev_tra	List with 10 folds of the train data from the EU>

```
Console
> library(mldr.datasets)
> data(package="mldr.dat
>
> mldrs()
>
```

Fig. 9.1 Looking at the available MLDs in the mldr.datasets package

To load any of the available MLDs, all the user has to do is typing in the R console its name followed by empty parentheses. The package will check whether the requested MLD is locally available into the user's computer, loading it into memory if this is the case. On the contrary, the MLD will be automatically downloaded from the RUMDR repository, stored in the local machine, and then loaded into memory, without needing any user intervention.

9.2.3.2   Exploring Loaded MLDs

The MLDs supplied by the mldr.datasets package are mldr objects. It is the object format defined by the mldr package, further addressed in this chapter. These objects have several members containing data helpful to explore the MLD structure, such as the names and frequencies of labels and labelsets and domains of input attributes. To access any of these members, the dataset$member syntax will be used, as depicted in Fig. 9.2.

The multilabel data are stored into the dataset member. This is a standard R data.frame; therefore, the usual R syntax to access any of its columns and rows is used. The measures() function returns a list of characterization metrics, such as the number of instances, features, and labels, imbalance levels, and theoretical complexity level.

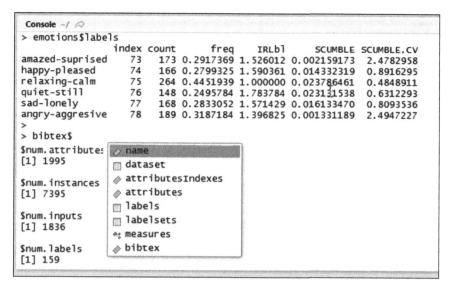

Fig. 9.2 The mldr objects have several members with disparate information

9.2.3.3 Obtaining Citation Information

When an MLD produced by any third party is going to be used in a new study, including the proper citation to give the original author, the correct attribution is mandatory. Obtaining the precise citation information is not always easy. The mldr.datasets package includes a custom version of the R's `toBibtex()` function whose goal is to provide the BibTeX entry associated with any `mldr` object.

The value returned by the `toBibtex()` function is properly formatted to copy it to the clipboard and then pasting it in the user's BibTeX editor. As demonstrated in Fig. 9.3, it can also be printed to the console.

9.2.3.4 Partitioning the MLDs

Although partitioned MLDs can be obtained from some repositories, this is not always the case. Furthermore, the number of partitions or their proportions could be not adequate for the user needs. The mldr.datasets package contributes two functions, named `random.kfolds()` and `stratified.kfolds()`, whose goal is to partition any `mldr` object into the number of desired parts. The difference between these two functions relies on the approach followed to choose the instances included in each partition. The former does it randomly, while the latter stratifies the data trying to balance the label distribution among partitions.

Both functions need the `mldr` object to be partitioned as their first argument. Additionally, they can take two more parameters specifying the number of folds, it

```
Console ~/                                                              ─□
> toBibtex(emotions)
[1] "@incollection{,\n  title = \"Multi-Label Classification of Emotions in Music\",\n  author = \"
Wieczorkowska, A. and Synak, P. and Ra'{s}, Z.\",\n  booktitle = \"Intelligent Information Processi
ng and Web Mining\",\n  year = \"2006\",\n  volume = \"35\",\n  chapter = \"30\",\n  pages = \"307-
-315\"\n}"
>
> cat(toBibtex(stackex_chess))
@inproceedings{,
  title="QUINTA: A question tagging assistant to improve the answering ratio in electronic forums",
  author="Charte, Francisco and Rivera, Antonio J. and del Jesus, Maria J. and Herrera, Francisco",
  booktitle="EUROCON 2015 - International Conference on Computer as a Tool (EUROCON), IEEE",
  year="2015",
  pages="1-6",
  month="sept"
}
> |
                                                                              I
```

Fig. 9.3 Obtaining the BibTeX entry to cite the MLD

```
Console ~/                                                              ─□
> randomFolds <- random.kfolds(emotions, k = 10)
> stratiFolds <- stratified.kfolds(emotions, k = 10)
>
> summary(randomFolds[[1]]$train)
  num.attributes num.instances num.inputs num.labels num.labelsets num.single.labelsets
1             78           534         72          6            26                     3
  max.frequency cardinality   density    meanIR   scumble scumble.cv      tcs
1            74    1.859551 0.3099251  1.479876 0.01131987   1.371573 9.326522
> summary(stratiFolds[[1]]$train)
  num.attributes num.instances num.inputs num.labels num.labelsets num.single.labelsets
1             78           533         72          6            25                     2
  max.frequency cardinality  density   meanIR   scumble scumble.cv      tcs
1            74    1.866792 0.311132 1.466343 0.01089432   1.286576 9.287301
> |
```

Fig. 9.4 An MLD being partitioned using random and stratified approaches

is 5 by default, and the seed for the random generator. The result returned by these functions is a list containing as many elements as folds have been indicated. Each one of these elements is made up of two members, called `train` and `test`, with the corresponding data partitions.

The example code shown in Fig. 9.4 demonstrates how to use these two functions, as well as how to access the training partition of the first fold. The `summary()` function prints a summary of characterization metrics, allowing to compare how the different partitioning approach has influenced the obtained partition.

9.2.3.5   Converting MLDs to Other Formats

Although R is a tool from which the MLDs provided by mldr.datasets can be used with disparate machine learning algorithms, currently software packages such as

MULAN and MEKA are preferred, due to the large prebuilt set of MLC algorithms they incorporate.

The file format of MLDs provided by RUMDR is the R native object format. Nonetheless, once they have been loaded into R, it is possible to convert them to several other file formats. For doing so, the `write.mldr()` function of the mldr.datasets package has to be called.

The `write.mldr()` function accepts as first argument an `mldr` object, containing the MLD to be written to a file. It is also able to deal with a list as the one returned by the partitioning functions described above, writing each training and test fold to a different file. This is the only mandatory parameter, and the remaining ones take default values.

As second argument, the `write.mldr()` function takes a vector of strings stating the file formats the data are going to be exported to. Valid formats are MULAN, MEKA, KEEL, LIBSVM, and CSV. The default value is `c("MULAN," "MEKA")`, being these the two most popular multilabel file formats. If the MULAN format is chosen, the function will also write the corresponding XML file for the MLD. For the CSV format, an additional CSV file containing a list with label names is also created.

The third parameter has to be a boolean value, indicating if sparse representation has to be used to write the data. By default, it takes the FALSE value, so the non-sparse format is chosen unless otherwise specified.

Lastly, the fourth argument sets the base filename the `write.mldr()` function will use to name the written files. This filename will be followed by a set of numbers, stating the fold and total number of folds, if the first parameter is a list of `mldr` objects.

The `write.mldr()` function can be combined with the previously described partitioning functions, as shown in Fig. 9.5. In this example, the yeast MLD is being partitioned into fivefolds, and then, the resulting partitions are written in MEKA and CSV file formats.

```
Console ~/
> write.mldr(stratified.kfolds(yeast), format = c('MEKA', 'CSV'), basename = 'yeast')
Wrote file yeast-1x5-tra.arff
Wrote file yeast-1x5-tra.csv
Wrote file yeast-1x5-tra_labels.csv
Wrote file yeast-1x5-test.arff
Wrote file yeast-1x5-test.csv
Wrote file yeast-1x5-test_labels.csv
Wrote file yeast-2x5-tra.arff
Wrote file yeast-2x5-tra.csv
Wrote file yeast-2x5-tra_labels.csv
Wrote file yeast-2x5-test.arff
Wrote file yeast-2x5-test.csv
Wrote file yeast-2x5-test_labels.csv
Wrote file yeast-3x5-tra.arff
Wrote file yeast-3x5-tra.csv
Wrote file yeast-3x5-tra_labels.csv
Wrote file yeast-3x5-test.arff
Wrote file yeast-3x5-test.csv
Wrote file yeast-3x5-test_labels.csv
Wrote file yeast-4x5-tra.arff
Wrote file yeast-4x5-tra.csv
wrote file yeast-4x5-tra_labels.csv
```

Fig. 9.5 Partitioning and exporting an MLD to MEKA and CSV file formats

Since due to its link to the RUMDR repository, most of the MLDs publicly available nowadays can be downloaded and then properly cited, partitioned, and exported to the common file formats, the mldr.datasets package can be the most convenient way of dealing with existent MLDs.

9.2.4 Generating Synthetic MLDs

The MLDs provided in the previous repositories have been generated from real data coming from different domains, as was explained in Chap. 3. However, sometimes having MLDs with very definite traits would be desirable, for instance while analyzing the behavior of algorithms aimed to solve a very specific problem. In order to produce these kinds of data, an appropriate algorithm has to be designed, usually including a mathematical method to correlate the inputs with the outputs.

In [12], such an algorithm, which offers two different approaches based on hypercubes and hyperspheres, is presented. The associated synthetic dataset generator for multilabel learning is available online, as a Web application (see Fig. 9.6). It allows the user to choose among the two strategies for generating instances, as well as to indicate how many features, labels, and instances the MLD will have, the noise level to add to the data, and other handful of configuration parameters. Once all of them have been set, the resulting dataset characteristics are shown and the file can be downloaded.

Mldatagen is a quite generic multilabel data generator. In the multilabel literature, some authors have created their own synthetic MLDs following more ad hoc approaches, adapted to suit specific goals. References to several of them can be found in [13].

Multilabel datasets are scattered through a collection of data repositories using disparate file formats. Once we know how to get these datasets and which are the file formats they use, a software tool such as the described mldr.datasets package is all we need to cite, partition, and export all MLDs in the proper format for the experiments we intend to do.

9.3 Exploratory Analysis of MLDs

The Web repositories where the MLDs are downloaded from, such as the previously mentioned MULAN and MEKA, usually also supply some basic information about the datasets. The number of instances, input attributes, and labels, along with label cardinality and sometimes label density, are common metadata. However, before using these MLDs to conduct some experiments, most users will demand additional details. Those will be generally obtained by means of exploratory analysis tasks, including summarizing at different levels and visualizing the data.

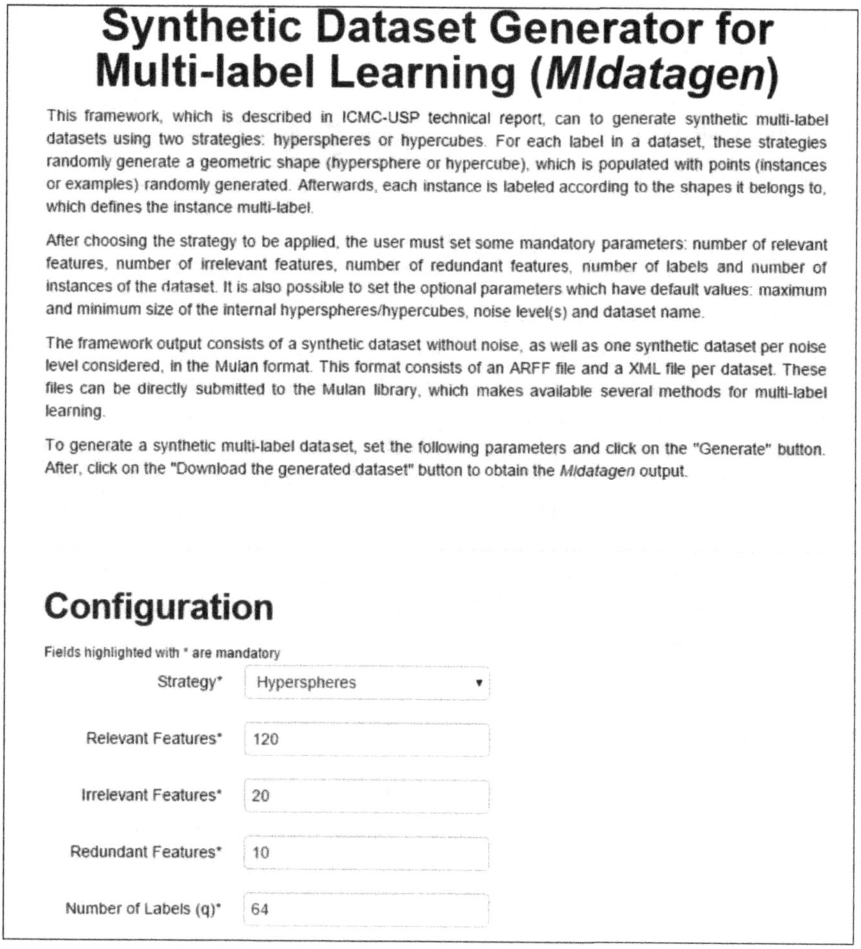

Synthetic Dataset Generator for Multi-label Learning (*Mldatagen*)

This framework, which is described in ICMC-USP technical report, can to generate synthetic multi-label datasets using two strategies: hyperspheres or hypercubes. For each label in a dataset, these strategies randomly generate a geometric shape (hypersphere or hypercube), which is populated with points (instances or examples) randomly generated. Afterwards, each instance is labeled according to the shapes it belongs to, which defines the instance multi-label.

After choosing the strategy to be applied, the user must set some mandatory parameters: number of relevant features, number of irrelevant features, number of redundant features, number of labels and number of instances of the dataset. It is also possible to set the optional parameters which have default values: maximum and minimum size of the internal hyperspheres/hypercubes, noise level(s) and dataset name.

The framework output consists of a synthetic dataset without noise, as well as one synthetic dataset per noise level considered, in the Mulan format. This format consists of an ARFF file and a XML file per dataset. These files can be directly submitted to the Mulan library, which makes available several methods for multi-label learning.

To generate a synthetic multi-label dataset, set the following parameters and click on the "Generate" button. After, click on the "Download the generated dataset" button to obtain the *Mldatagen* output.

Configuration

Fields highlighted with * are mandatory

Strategy*	Hyperspheres ▾
Relevant Features*	120
Irrelevant Features*	20
Redundant Features*	10
Number of Labels (q)*	64

Fig. 9.6 The Mldatagen tool allows creating custom synthetic MLDs

As long as the structure of each multilabel file format is adequately decoded, any generic data manipulation tool could be used to explore the MLDs. Nevertheless, this section is focused on interactive software tools specifically built to work with multi-label data. Two particular tools in this category are described below, MEKA and the mldr package. Both provide some EDA (*Exploratory Data Analysis*) functionality.

9.3.1 MEKA

MEKA is a software tool built upon WEKA, and it brings a similar user interface to this popular application but with multilabel capabilities. MEKA is developed in Java; therefore, to use it, the first requirement to meet is having the Java Runtime Envi-

Fig. 9.7 The MEKA main window allows the user to open several tools. The Explorer lets open and explore MEKA datasets, as well as to interactively perform experiments with them

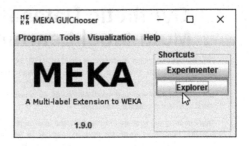

ronment (JRE) installed in the system. Then, the most recent version of MEKA can be downloaded from https://adams.cms.waikato.ac.nz/snapshots/meka. The installation process is just extracting the files of the compressed file to a folder.

In addition to the software itself, the MEKA compressed file also includes a PDF document with a tutorial and some example data files. Two scripts, one for Windows (run.bat) and another one for UNIX-based systems run.sh, aimed to ease the launch of the software are provided as well.

Launching MEKA through the proper script will open the program's main window (see Fig. 9.7). The options in the **Tools** menu run the essential MEKA tools. Some of them are depicted below.

9.3.1.1 The ARFF Viewer

MEKA includes a viewer able to open any ARFF file, enumerating its attributes, including the labels, and the values assigned to them. Once a dataset has been loaded, its content is shown in a grid like the one in Fig. 9.8. The **Properties** option in the **File**

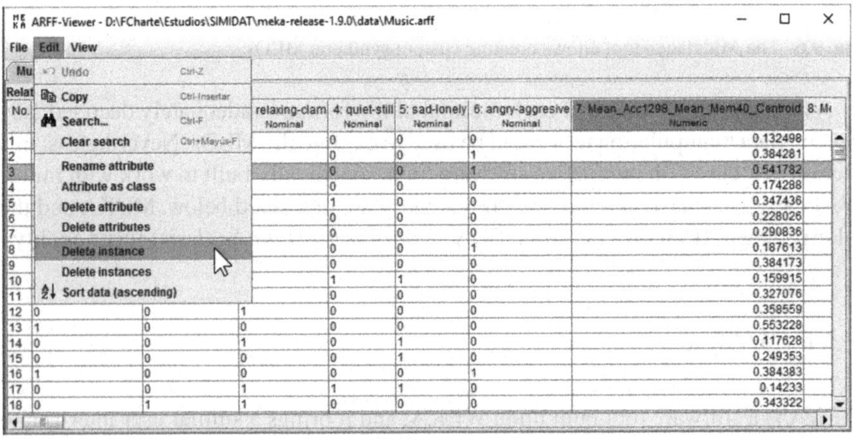

Fig. 9.8 The MEKA ARFF viewer allows viewing and editing any ARFF dataset contents

menu will report the number of instances, attributes, and classes. This is a generic ARFF tool, so it is not aware of the multilabel nature of the data.

Actually, this tool is also an editor, since the values can be changed, instances and attributes can be removed, and the order of the data samples can be modified. Any change will be lost unless the data in memory are saved, usually with a different filename.

9.3.1.2 Attribute Exploration

The other exploration tool embedded in MEKA is the MEKA Explorer, accessible through the **Tool**→**Explorer** option in the main window. This tool is aimed to interactively test preprocessing and classification algorithms, but it also has some exploration capabilities.

After loading an MLD, the **Preprocess** page will show two lists containing all the attributes in the dataset (see Fig. 9.9). At the top of the left list, there is a summary with the number of instances and attributes. If the MLD has been loaded from a file in MEKA format, the program will automatically detect which attributes are labels. These will be highlighted in bold and always will be at the beginning in both lists. Loading MLDs from other ARFF-based file formats is allowed, but the program will be not able to identify the labels. The user has to mark them in the right list and then

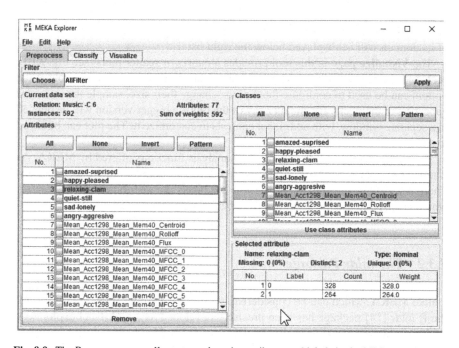

Fig. 9.9 The Preprocess page allows to explore the attributes and labels in the MLD

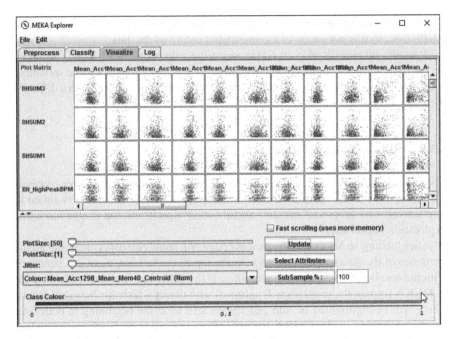

Fig. 9.10 Pairwise scatter plots for the attributes in the MLD

click the **Use class attributes**. By selecting any attribute in the left list, a summary of its domain will be displayed below the right list.

The **Visualize** page in this tool provides a matrix of scatter plots (see Fig. 9.10), each one showing the relationship between a pair of attributes. The controls at the bottom can be used to configure these plots, change their size, sample the instances appearing in them, etc. By clicking any of the plots, a bigger version will be opened in a separate window, with additional controls to customize it. The plot can be saved to a file.

Basically, these are all the exploratory analysis functions offered by MEKA. As can be seen, they are mostly applicable to both traditional and multilabel datasets. There is a lack of information about how the labels are distributed and correlated, which labelsets and other specific multilabel measurements exist.

9.3.2 The mldr Package

Unlike MEKA, which can be considered as a general-purpose multilabel software, the R mldr [14] package has been precisely developed as a multilabel exploratory analysis tool. This package is also available at CRAN, like the previously described mldr.datasets package, so it can be installed in the same way. There are not dependencies between the two packages. That means that mldr can be used without installing mldr.datasets and vice versa. Nonetheless, mldr can take advantage of having all the MLDs included in the mldr.datasets package, as well as their functions to partition and export them.

The mldr package is able to load MLDs from MULAN and MEKA file formats, as well as to generate new MLDs on the fly from data synthetically generated or provided by the user in a `data.frame`. The package defines a custom representation for multilabel data. The MLDs are R S3 objects with `mldr` class. This is the same representation used by the MLDs in the mldr.datasets package, and hence, these datasets are compatible with mldr.

Inside the mldr package, a plethora of characterization metrics are computed, along with a set of functions aimed to ease the analysis of multilabel data. In addition, the package provides a Web-based user interface to speed up exploratory tasks, including specific graphic representations of the data.

In this section, the procedures to accomplish different exploratory duties using the mldr package are explained. It is assumed the user has installed the package, by issuing the `install.packages("mldr")` command at the R console, and it is loaded into memory, by typing the `library(mldr)` command.

9.3.2.1 Loading and Creating MLDs

After loading the mldr package, three `mldr` objects will be already available. These correspond to the `birds`, `emotions`, and `genbase` MLDs. If the mldr.datasets is also loaded, all the MLDs provided by it will be accessible as well. To load any other MLD, assuming it is stored in an ARFF file using the MEKA or MULAN formats, the `mldr()` function has to be called. The only mandatory argument is the filename without extension. The function will assume the MLD is in MULAN file format and the existence of an XML file with the same name. Additional parameters can be supplied to change this default behavior, stating the XML filename, the number, indexes, or names of the attributes acting as labels if there are not an XML file available nor this information is provided in the ARFF header, etc.

The `mldr()` function always checks whether the mldr.datasets package is installed in the system. If this is the case, the function entrusts the loading of the MLD to the proper mldr.datasets function. To avoid this functionality, forcing the loading from a local file, the `force_read_from_file` parameter has to be included in the call to `mldr()`, assigning it the `TRUE` value.

In Fig. 9.11, a use case of the `mldr()` function is shown. In this example, the name of the XML file does not coincide with the ARFF filename, so it is explicitly set by means of the `xml_file` parameter. Once the MLD has been loaded, the object can be queried to obtain some dataset traits as demonstrated.

Furthermore, new MLDs can be created from any existing data, whether it is from a real domain or produced by any formula. This functionality, provided by the `mldr_from_dataframe()` function, allows creating synthetic MLDs. This function needs as inputs a `data.frame` containing all the features, a vector stating

```
Console D:/FCharte/Estudios/Publicaciones/ML-SMOTE/AdditionalMaterial/
> yeast <- mldr('yeast-5x2x1-1tra', xml_file = 'yeast.xml')
> summary(yeast)
  num.attributes num.instances num.inputs num.labels num.labelsets
1             117          1933        103         14           174
            66           189    4.222969
  num.single.labelsets max.frequency cardinality

    density  meanIR scumble scumble.cv      tcs
1 0.3016407 7.41181 0.105536   1.046271 12.43284
> |
```

Fig. 9.11 Loading an MLD from an external ARFF file

```
Console D:/FCharte/Estudios/Publicaciones/ML-SMOTE/AdditionalMaterial/
> features <- data.frame(matrix(rnorm(500), ncol = 5))
> genLabels <- function() c(sample(c(0, 1), 100, replace = TRUE))
> features$labelA <- genLabels()                                      I
> features$labelB <- genLabels()
>
> syntheticMLD <- mldr_from_dataframe(
+     features, labelIndices = c(6, 7),
+     name = 'SyntheticMLD')
>
> summary(syntheticMLD)
  num.attributes num.instances num.inputs num.labels
1              7           100          5          2
  num.labelsets num.single.labelsets max.frequency cardinality
1             4                     0            37        0.79
    density   meanIR        scumble scumble.cv      tcs
1     0.395 1.097222 0.0006293422   2.302831 3.688879
> |
```

Fig. 9.12 New MLDs can be generated on the fly from any formula

which ones of them will be labels, and optionally a name to assign to the new MLD. The result, as is shown in Fig. 9.12, is an `mldr` object that can be used as any other MLD.

9.3.2.2 Querying Data Characterization Metrics

Independently of the origin of the MLD's data, all `mldr` objects have the same structure and can be processed with the same set of functions. Many of these are aimed to compute and provide several characterization metrics. The supplied metrics can be grouped into four categories:

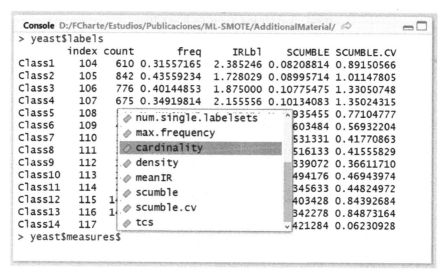

Fig. 9.13 A large set of traits are computed for each MLD

- **Basic traits**: The number of instances, number of input attributes, number of output labels, and number of labelsets are in this group. All of them can be queried with the syntax `mldrobject$measures$num.XXX`. The `summary()` function also returns this information.
- **Label distribution data**: Metrics such as label cardinality, label density, and the frequency of each individual label are available through the `measures` and `labels` members, as demonstrated in Fig. 9.13.
- **Label relationship metrics**: The relationships among the labels in the dataset can be inspected through metrics such as the total number of unique labelsets (`measures$num.labelsets`) in the MLD, the number of single labelsets (`measures$num.single.labelsets`), and analyzing the values for the global and by label *SCUMBLE* measures (`measures$scumble` and `labels$SCUMBLE`).
- **Metrics related to label imbalance**: The individual imbalance level for each label is provided by the `labels$IRLbl` member. The average imbalance level for the MLD can be obtained through the `measures$meanIR` member.

Additional information about the MLD is provided in the members of the `mldr` object such as `labelsets` (signature and counter of each labelset in the MLD), `attributes` (name and domain of each attribute), and `measures$tcs` (the theoretical complexity score of the MLD). All of them are properly documented in the electronic help included in the package.

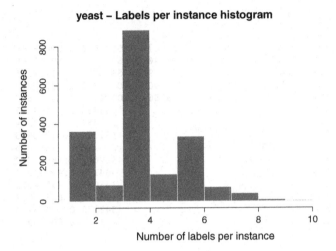

yeast – Labels per instance histogram

Fig. 9.14 Number of labels per instance histogram

9.3.2.3   mldr Custom Plots

The mldr package delivers a custom `plot()` function for `mldr` objects, able to produce seven specific plots from the data contained in those objects. The arguments to this function are usually two, the `mldr` object to analyze and the `type` parameter specifying what type of plot is desired. Some kinds of visualizations accept additional parameters, for instance to restrict the set of plotted labels.

Three of the custom plots are histograms designed to depict how labels and labelsets are distributed among the instances. The one shown in Fig. 9.14 is produced by the `plot(yeast, type = "CH")` call, showing the distribution of label cardinality among the instances. The `yeast` dataset has a *Card* value of 4.223. That most instances have four labels can be observed in this plot.

Two more of the available types, `"LB"` and `"LSB,"` are bar plots depicting how many instances each label and labelset appear. This way, the differences between frequencies of labels and labelsets can be explored. For instance, Fig. 9.15 shows that there are two majority labels (`Class12` and `Class13`) and several minority labels (`Class14` and `Class9` to `Class11`).

Another kind of plot is denoted as `"AT."` It is a regular pie chart showing the proportion of each type of features, continuous, nominal, and labels. The seventh visualization option is the one generated by default when the `type` argument is not provided. It is a circular plot like the one described in Chap. 8 (see Fig. 9.16), illustrating how the labels interact among them. This kind of plot, as well as those presenting label frequencies, accepts the optional parameters `labelCount` and `labelIndices`, whose goal is to restrict the set of drawn labels. All plot types also take other optional arguments, such as `title` to set the title of the plot and `col` to set the color, or the standard graphical parameters usually given to R functions such as `barplot()` and `hist()`.

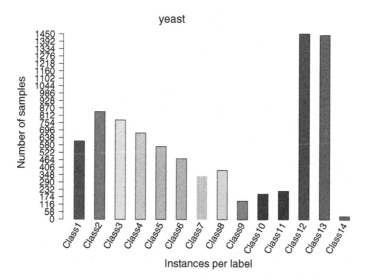

Fig. 9.15 The differences among label frequencies are easily inferred in this bar plot

9.3.2.4 Automated Reports

The mldr package includes some functions able to generate reports by means of automated analysis of the label measurements. The functions in charge of producing these reports only need the `mldr` object to be analyzed as argument. The resulting report is printed to the console.

With the `labelInteractions()` function, an analysis of which labels are in minority and how they interact with others is generated. The reported result includes the indexes of the minority labels, at the beginning, and for each one of them the list of labels they have interactions with, along with the number of samples in which they appear together (see Fig. 9.17).

The second report is produced by the `concurrenceReport()` function. As its name suggests, the report analyzes the concurrence among labels, but in a more elaborated way than the `labelInteractions()` function. As is shown in Fig. 9.18, the report includes the global *SCUMBLE* measurement and its CV, as well as *SCUMBLE* values for each individual label. For each minority label, a list of the labels it interacts with is also provided, including label names and indexes, the number of times they appear together, and their respective *SCUMBLE* values.

The information given by these functions can be useful in different scenarios. For instance, the indexes of minority labels and the labels they interact with would be convenient to customize the interaction plot previously described.

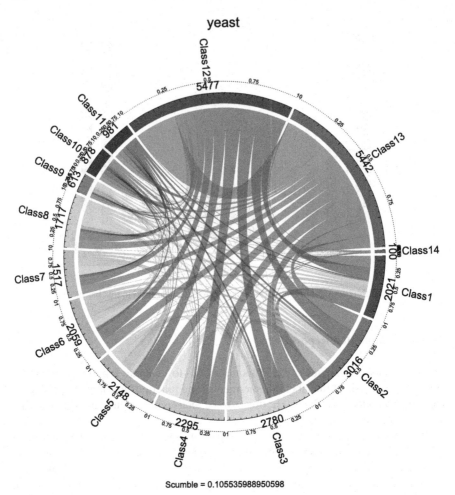

Fig. 9.16 Label interactions in the yeast MLD

9.3.2.5 The mldr User Interface

All the exploratory functions in the mldr package are accessible through the command line, so that R scripts can be written to perform reproducible analysis procedures. However, all the functionality already described is also reachable through the integrated GUI. To open it, the user has to enter the `mldrGUI()` sentence. The GUI will be launched inside the default Web browser, consisting of several sections.

Just after the GUI has been launched, the main section is shown and the first MLD available in the working environment is selected. The name of the active MLD is always shown at the top of the GUI. A panel in this section (see Fig. 9.19) allows the user to choose from the loaded MLDs, as well as to load any others from their files.

```
Console D:/FCharte/Estudios/Publicaciones/ML-SMOTE/AdditionalMaterial/
> labelInteractions(yeast)
$indexes
[1] 117 112

$interactions
$interactions$Class14

104 105 106 107 108 109 110 111 115 116
  1   2  26  26   6   1   1   2  17  17

$interactions$Class9

104 105 106 107 108 109 110 111 113 114 115 116
 42  77  62  18  35  33  23 121  41  10  75  75


>
```

Fig. 9.17 Report about how minority labels interact with other labels

```
Console D:/FCharte/Estudios/Publicaciones/ML-SMOTE/AdditionalMaterial/
> concurrenceReport(emotions)
Dataset musicout: Mean SCUMBLE 0.01095238 with CV 1.26456

SCUMBLE mean values by label:
# relaxing-calm: 0.02379
# quiet-still: 0.02313
# sad-lonely: 0.01613
# happy-pleased: 0.01433
# amazed-suprised: 0.002159
# angry-aggresive: 0.001331

Minority label quiet-still (76, SCUMBLE 0.02313154) interacts with:
# relaxing-calm (75, SCUMBLE 0.02378646): 104 interactions
# angry-aggresive (78, SCUMBLE 0.001331189): 2 interactions

Minority label sad-lonely (77, SCUMBLE 0.01613347) interacts with:
# relaxing-calm (75, SCUMBLE 0.02378646): 95 interactions
# angry-aggresive (78, SCUMBLE 0.001331189): 20 interactions

Minority label happy-pleased (74, SCUMBLE 0.01433232) interacts with:
# relaxing-calm (75, SCUMBLE 0.02378646): 91 interactions
# angry-aggresive (78, SCUMBLE 0.001331189): 12 interactions
```

Fig. 9.18 Report about concurrence of labels in the MLD

A visual summary of the selected MLD is provided at the right, in the same section, including plots that show the proportion of each kind of attributes and how labels and labelsets are distributed. These plots can be saved for further use, for instance including them in any study.

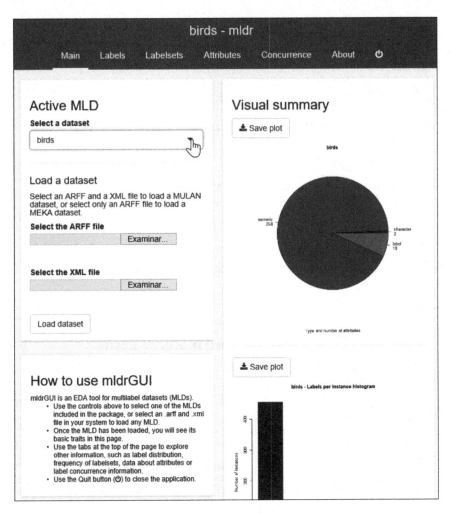

Fig. 9.19 Main page of mldr's GUI

The tabs located at the top of the user interface, below the selected MLD name, are the doors to the other GUI sections. The **Labels** and **Labelsets** pages are quite similar. Both provide a table with names and frequencies, as well as the associated bar plot. The labels table includes additional data, such as the feature index, imbalance levels, and concurrence levels. These tables can be sorted and filtered, and their content can be printed and exported, as is shown in Fig. 9.20.

In the **Attributes** page, the domain of each input attribute is summarized. The frequency of each value is computed for nominal attributes, while for numeric ones, some statistical measures are computed (see Fig. 9.21).

By opening the **Concurrence** section, the user can access a concurrence analysis report, along with a customizable circular plot. The labels presented in it can be inter-

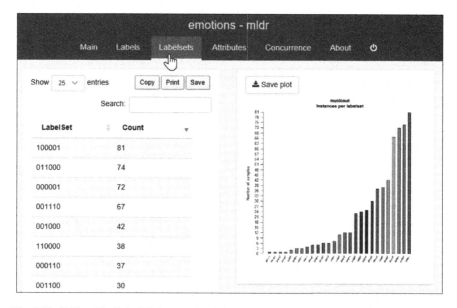

Fig. 9.20 Table with all the labelsets and their frequencies and the corresponding plot

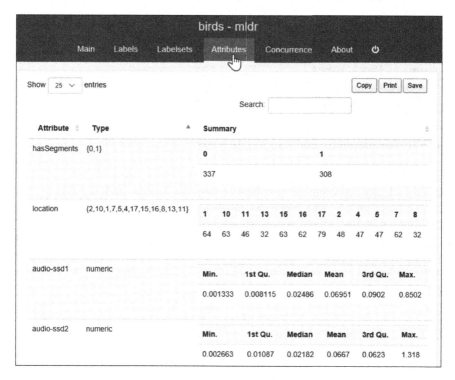

Fig. 9.21 The Attributes page provides a summary of each attribute domain

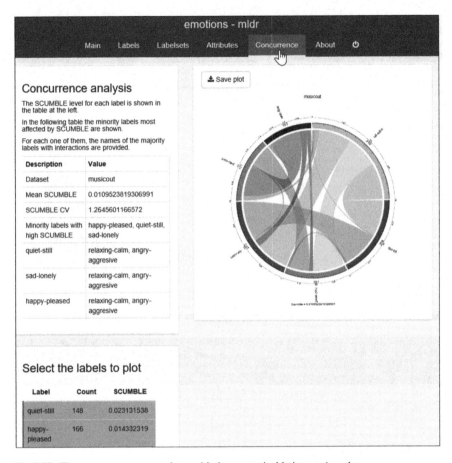

Fig. 9.22 The concurrence report along with the customizable interaction plot

actively selected, updating the visualization until the desired result is achieved. The report is based on the concurrence and imbalance metrics, indicating which minority labels have high interactions and which labels interact with others (Fig. 9.22).

Overall, the exploratory functions and GUI of the mldr package supply the user with an extensive range of useful information. In addition, the GUI allows customizing the tables and plots, and then exporting them, easing the process of documenting any report or study.

9.3.2.6   Other mldr Functions

In addition to the exploratory analysis functionality, the mldr package also offers the user functions to transform and filter the MLDs, as well as to evaluate the predictive performance from the outputs obtained by any classifier. Some of these functions are

```
Console  D:/FCharte/Estudios/Publicaciones/ML-SMOTE/AdditionalMaterial/
>
> yeast[Class13 == 0]$measures$num.instances
[1] 489
>
> summary(yeast[Class14 == 1] + yeast[.SCUMBLE > 0.2])
  num.attributes num.instances num.inputs num.labels num.labelsets
1            117           342        103         14            73
  num.single.labelsets max.frequency cardinality   density
1                   29            61    5.391813 0.3851295
    meanIR    scumble scumble.cv      tcs
1 2.959099 0.1121697  0.4175117 11.56425
>
```

Fig. 9.23 Filtering data instances and joining two subsets

implemented as operators, preserving the natural syntax used in the R language to accomplish similar operations.

By means of the == and + operators, two mldr objects can be compared and combined. Two MLDs are equal as long as they have the same structure and content. To join two MLDs, they have to be structurally identical. With the [ ],[3] the instances in the MLD can be filtered according to any valid expression. An example on how to use the latter operator to get the samples in which a certain label appears is shown in Fig. 9.23. The result returned by the operator is also an mldr object; thus, it can be used as any other mldr dataset. The same is applicable to the + operator.

The basic BR and LP multilabel data transformations can be applied to any mldr object through the mldr_transform() function. The only arguments needed are the MLD to transform and the type parameter indicating which transformation to apply. The valid values are "BR" and "LP." Optionally, a third parameter can be given stating the labels to use. This way, the transformation can be limited to a subset of labels. The value returned by this function will depend on the kind of transformation requested. For BR, it will be a list with as many data.frame objects as labels, each containing a binary dataset. For LP, only a data.frame is produced as output.

As is shown in Fig. 9.24, a column named classLabel is introduced in the data.frame instead of the original labels. This way, the resulting data.frame can be used with any of the binary or multiclass classifiers available in R, using classLabel as the class to predict.

Although the mldr package does not provide any multilabel classifier, a function to evaluate predictions obtained by any other means is supplied. This function, named mldr_evaluate(), takes two arguments, the mldr object with the instances

---

[3]The [ ] operator defined in the mldr package is designed to work with mldr objects only. The standard [ ] R operator can be used over the mldr$dataset member to manipulate the raw multilabel data.

```
Console  D:/FCharte/Estudios/Publicaciones/ML-SMOTE/AdditionalMaterial/
>
> yeastBR <- mldr_transform(yeast, type = "BR")
> class(yeastBR)
[1] "list"
> length(yeastBR)
[1] 14
>
> yeastLP <- mldr_transform(yeast, type = "LP")
> class(yeastLP)
[1] "data.frame"
>
> head(yeastLP$classLabel)
[1] 11000000000110 01100000000110 00111100000110 01100011000110
[5] 00000000011110 00110000000110
174 Levels: 00000000000110 00000000001100 ... 11111001100110
>
```

Fig. 9.24 The mldr package provides a function to perform the usual BR and LP transformations

being assessed and the predictions obtained for them. The function returns a list containing about twenty performance metrics (see Fig. 9.25), along with an object containing all the data needed to plot the ROC curve. The example shown in Fig. 9.25 generates random predictions for all the labels in the yeast dataset and then evaluates the performance. As expected, the accuracy of this random classifier is around 50 %.

```
Console  D:/FCharte/Estudios/Publicaciones/ML-SMOTE/AdditionalMaterial/
>
> pred <- matrix(sample(0:1, yeast$measures$num.instances*yeast$measu
res$num.labels, replace = TRUE), ncol = yeast$measures$num.labels)
>
> str(mldr_evaluate(yeast, pred))
List of 20
 $ Accuracy        : num 0.507                    I
 $ AUC             : num 0.606
 $ AveragePrecision: num 0.428
 $ Coverage        : num 9.67
 $ FMeasure        : num 0.404
 $ HammingLoss     : num 0.493
 $ MacroAUC        : num 0.514
 $ MacroFMeasure   : num 0.343
 $ MacroPrecision  : num 0.308
 $ MacroRecall     : num 0.519
 $ MicroAUC        : num 0.51
 $ MicroFMeasure   : num 0.387
 $ MicroPrecision  : num 0.31
 $ MicroRecall     : num 0.517
 $ OneError        : num 0.671
 $ Precision       : num 0.311
 $ RankingLoss     : num 0.0745
```

Fig. 9.25 If a set of predictions is available, the performance of the classifier can be assessed

Exploratory analysis is a fundamental step to understand the data we are working on, providing the necessary knowledge to decide which preprocessing and learning methods should be applied. MEKA offers a GUI with some general tools, based on the popular WEKA. On the other hand, the mldr package delivers a plethora of specific multilabel metrics, plots, and reports, along with several functions to manipulate these kinds of data.

9.4 Conducting Multilabel Experiments

Once the user has learned what the structure of their MLDs is, how labels are distributed and correlated, etc., it is time to conduct some experiments using the multilabel data. This goal can imply applying some preprocessing method to the data, train a classifier using it, obtain predictions for the test set, and eventually evaluate these predictions to assess the performance.

Several software tools and packages for different languages are available to accomplish these tasks. However, there are two applications that stand out among everything else, MEKA and MULAN. Both have been developed by experts on the multilabel field, and they provide reference implementations for a large variety of MLC algorithms.

In this final section, how to use MEKA and MULAN to run simple multilabel experiments is explained. Although the former tool can be used from the command line, it is supplied with a GUI which eases the user's work. On the contrary, the latter only is accessible programmatically, as will be later shown. Furthermore, both software packages have been programmed in Java language, so that the latest JRE installed in the system is assumed.

9.4.1 MEKA

The MEKA user interface was previously introduced. Specifically, the tool known as MEKA Explorer was used to make some exploratory analysis on the data. This section starts from this earlier knowledge, firstly describing how to conduct interactive experiments and then designing more complex ones.

9.4.1.1 Running Experiments Interactively

The **Classify** page on this application allows the user to choose among an extensive list of classifiers, training them with the loaded MLD and evaluating them following

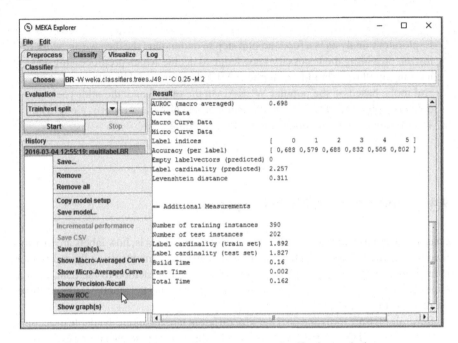

Fig. 9.26 The MEKA Explorer allows the user to run any classifier interactively

several validation strategies. The first step is to load the MLD of interest. Then, by means of the **Choose** button in the **Classify** page, the classifier is selected, setting its parameters as desired. The drop-down list in the **Evaluation** section configures the validation scheme.

Once the experiment has been configured, it will be run by clicking the **Start** button. As soon as it finishes displaying a summary of results, as is shown in Fig. 9.26. Each experiment execution is independently stored in the **History** list, so the user can compare the results produced in each case.

The pop-up menu associated with the items in the **History** list can be used to save the obtained results, delete the experiment, copy its parameters, and show several performance evaluation plots. This menu appears opened in Fig. 9.26. By selecting the **Show ROC** option, a new window like the one shown in Fig. 9.27 is opened. It contains several pages, one per label, each one with the ROC plot.

Overall, the MEKA Explorer offers the user with a very simple way to run individual experiments. However, they are limited to using one MLC algorithm over one MLD.

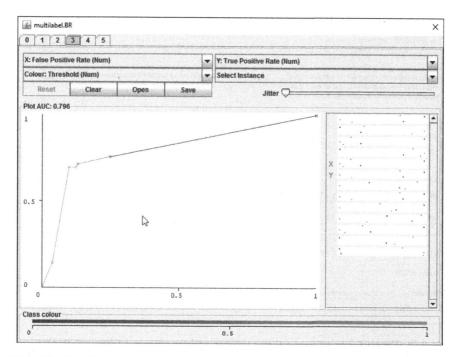

Fig. 9.27 The ROC curve for the selected experiment is shown in a window in its own

9.4.1.2 Designing Complex Experiments

Though running individual MLC algorithms with one MLD can be useful many times, for instance while analyzing the algorithm behavior, a real multilabel experimentation usually implies applying several methods to a group of MLDs. The MEKA Experimenter, another tool accessible from the MEKA main window, better suits this kind of job. As the MEKA Explorer, the Experimenter user interface also consists of several pages. The first one is where the user will design the experiment, fulfilling the following steps:

1. Adding the MLC algorithms to be used to the left list. The buttons at the right of this list allow the user to add and remove methods, configure them, and change the order they will be run.
2. Adding the MLDs which the algorithms will be applied to the right list. The buttons at the right of this list are similar to the previous ones.
3. Configuring how many times the experiment will be run, as well as how the results will be evaluated. For instance, in Fig. 9.28, a configuration with 10-fold cross-validation and 10 repetitions has been established.
4. Launching the experimentation by choosing the **Start** options in the **Execution** menu. Previously, all the configuration can be saved.

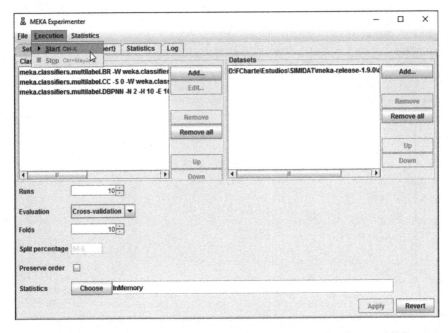

Fig. 9.28 The MEKA Experimenter is able to run several classifiers using disparate MLDs

Once the batch of runs defined in the experimentation finishes, the obtained evaluation results can be analyzed through the options in the **Statistics** page of the MEKA Experimenter. An extensive set of performance metrics is provided, and they can be viewed separately or aggregating them. For instance, it is possible to choose a specific metric and get the average values for each algorithm and dataset. This way, a comparative study can be easily performed.

The results produced by the experiments can be also written to a file, whether the user is interested in raw or aggregated data. As shown in Fig. 9.29, the options relating to exporting functions can be found in the **Statistics** menu.

9.4.2   MULAN

As MEKA, MULAN [1] is a multilabel software framework built on top of WEKA. However, it does not bring the user with a GUI. All the tasks have to be accomplished by writing Java code. Therefore, some experience with this programming language is essential. In addition to the aforementioned JRE, to use MULAN the Java Development Kit (JDK) is also needed. The JDK contains the compiler, among other Java tools and utilities.

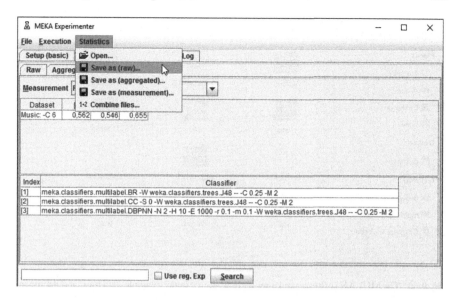

Fig. 9.29 The experimental results can be viewed in MEKA Experimenter and exported to a file

In this section, the procedures to design a multilabel experiment using MULAN are briefly described. Additional example code can be found both in the MULAN Web site (http://mulan.sourceforge.net) and in this book's GitHub repository.

9.4.2.1 Obtaining MULAN

The first step of a MULAN user is obtaining the software itself. The last version can be downloaded from http://mulan.sourceforge.net/download.html as a single compressed file. After decompressing it, a folder named `mulan` will be found, containing the folders and files shown in Fig. 9.30. Nearly all the folder names are self-explanatory. The Java JAR package holding the MULAN classes is in the `dist` folder, while the WEKA package needed to run MULAN is provided in the `lib` folder.

9.4.2.2 Compiling a MULAN Experiment

Once MULAN is installed in the user system, to conduct any experiment a Java source file has to be created. It will contain all the code to load the MLDs, to select and configure the MLC algorithms to run, to evaluate the obtained results, etc. These tasks imply using some MULAN classes; therefore, the proper dependencies have to be imported at the beginning of the source code. Assuming the user needs to load an MLD and wants to use the ML-kNN classifier, the following sentences will import the corresponding classes:

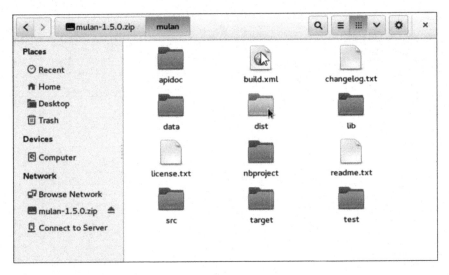

Fig. 9.30 The MULAN software packages include source code, example data files, and documentation

```
import mulan.data.MultiLabelInstances;
import mulan.classifier.lazy.MLkNN;
```

The way some of these MULAN classes are used is a matter which will be further described. Considering the code is already written and stored in a file named `MulanExperiment.java`, the compilation process involves calling the Java compiler providing the path to libraries and the aforementioned filename. The example in Fig. 9.31 shows a GNU/Linux terminal with the command line used to compile this hypothetical program. After compilation, a `.class` file with the compiled version is generated. To run this program, a similar command would be used, but changing `javac` for `java`.

```
                          francisco@dmserver: ~/mulan                              ✕

File  Edit  View  Search  Terminal  Help
fcharte@MULAN$
fcharte@MULAN$ javac -cp lib/weka-3.7.10.jar:dist/mulan.jar MulanExperiment.java

fcharte@MULAN$ ls -l MulanExperiment.*
-rw-rw-r-- 1 francisco francisco 1049 mar  1 21:43 MulanExperiment.class
-rw-rw-r-- 1 francisco francisco  689 mar  1 21:39 MulanExperiment.java
fcharte@MULAN$ █
```

Fig. 9.31 To compile a MULAN program, the paths to the mulan.jar and weka.jar files have to be provided

9.4.2.3 Loading Data Files

The `MultiLabelInstances` class is able to load a set of multilabel instances from a file. This class constructor usually is given two parameters, the path of the ARFF file and the path of the associated XML file. Other ways to obtain the data are considered, for instance from existing WEKA data samples and an array stating which ones are labels.

An initialized `MultiLabelInstances` object can be passed as argument to different methods. It also offers several functions which return data traits, such as the number of instances or labels, and label cardinality.

Assuming the user is working in the `mulan` folder, so a `data` subfolder with some example data is available, and the code below will load the emotions dataset and then print the number of instances and labels and the label cardinality. Figure 9.32 shows how the program is compiled and executed.

```java
import mulan.data.MultiLabelInstances;

public class MulanExperiment {
  public static void main(String[] args)
  throws Exception {
    MultiLabelInstances emotions =
      new MultiLabelInstances("data/emotions.arff",
                              "data/emotions.xml");

    System.out.println(
        "\nInstances:" + emotions.getNumInstances() +
        "\nLabels:" + emotions.getNumLabels() +
        "\nCardinality:" + emotions.getCardinality());
  }
}
```

```
                    francisco@dmserver: ~/mulan                              x

 File  Edit  View  Search  Terminal  Help
fcharte@MULAN$ javac -cp lib/weka-3.7.10.jar:dist/mulan.jar MulanExperiment.java

fcharte@MULAN$ java -cp lib/weka-3.7.10.jar:dist/mulan.jar:. MulanExperiment

Instances: 593
Labels: 6
Cardinality: 1.8684654300168635
fcharte@MULAN$ ▮
```

Fig. 9.32 The program loads the MLD and outputs some data about it to the console

The name of the files to load could obviously be supplied in the command line, instead of being hardwired in the Java code as in this simple example.

9.4.2.4   Configuring MLC Algorithms

Once the data are already in a `MultiLabelInstances` object, the next step is to configure the MLC algorithms this object is going to be given as input. MULAN provides a large set of classification algorithms, they are held in the `mulan.classifier` namespace, and some partitioning and preprocessing methods spread out several namespaces. All of them can be easily found in the electronic help of the program.

Many of the algorithms included in MULAN can work using default values, so the corresponding object is created without needing any parameters. For instance, to work with the ML-kNN algorithm (see Chap. 5), all the user has to do is to create an `MLkNN` object, as follows:

```
MLkNN kNNClassifier = new MLkNN();
```

To change the default values for the algorithm, a different constructor accepting them should be called. Other classifiers, such as the ones based on transformation methods, always need at least one parameter, specifically the base binary or multiclass method to be used as underlying classifier. Any classifier available as a WEKA class can be in charge of this task.

The following sentences would create two classifier instances. The first one is ML-kNN with 5 nearest neighbors, while the second one is a BR transformation using the standard C4.5[4] algorithm as base classifier. The last sentence prints in the standard output the basic information about the ML-kNN algorithm.

```
MLkNN kNNClassifier = new MLkNN(5, 1);
BinaryRelevance BRClassifier =
    new BinaryRelevance(new J48());

System.out.println(
    "\nkNNClassifier:" +
    kNNClassifier.getTechnicalInformation().toString());
}
```

9.4.2.5   Training and Evaluating the Classifier

MULAN has a class named `Evaluator` able to take a full dataset, partition it, and conduct a cross-validation evaluation. The method to accomplish the full procedure

---

[4]The C4.5 algorithm is implemented in WEKA by the `J48` class.

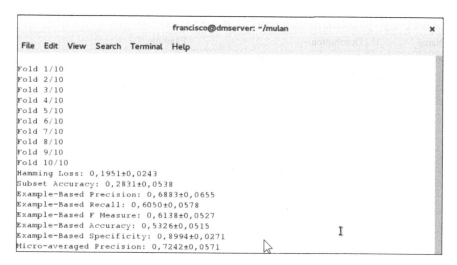

Fig. 9.33 Partial output produced by printing the information returned by crossValidate()

is crossValidate(), and it needs three parameters: the classification model, the dataset, and the number of folds to be used.

Assuming the proper namespaces have been imported and the emotions variable is a MultiLabelInstances object with the MLD, the following sentence would produce an output similar to that shown in Fig. 9.33. A 10-fold cross-validation is performed using the ML-kNN algorithm, and average values for a large set of evaluation metrics are returned.

```
System.out.println(
    new Evaluator().crossValidate(
        new MLkNN(), emotions, 10));
```

Instead of relying on an automated partitioning, training, and evaluation process, the users can separately run each step on their own. The training and test partitions, maybe generated by the functions in the mldr.datasets package previously described, can be individually loaded from their files. The full dataset can be also partitioned using the IterativeStratification class. In any case, two or more MultiLabelInstances objects will be used.

The training of any MULAN classifier is accomplished by calling its build() method. It needs only one argument, and the MultiLabelInstances objects with the samples aimed to train the model. Once trained, the model can be evaluated or used to get predictions for new instances. The former task is handled by the evaluate() method of the Evaluator class, taking as input the model and the test instances. The latter is in charge of the makePrediction() method of the model itself. It takes a MultiLabelInstance object as input, with the instance the prediction is aimed for.

Table 9.1 Input parameters for the RunMLClassifier utility

Name	Description	Example
`-path`	Establishes the path where the data files are located	`-path ~/data`
`-dataset`	Indicates the root name shared by all the dataset partitions. Training files have to be named as `MLD-strategy-Ntra.arff`, while test files will be `MLD-strategy-Ntst.arff`. The `strategy` part can be any user identifier for the partitioning strategy. `N` will be a sequential partition number. That an XML file with the name `MLD.xml` file is also in the folder is assumed	`-dataset emotions-10cv`
`-folds`	Sets the number of folds to iterate. The `N` part of the dataset name will take values from 1 to the number specified by this argument	`-folds 10`
`-algorithm`	Chooses the MLC algorithm to be used. The list of values accepted by this parameter is shown in Table 9.2	`-algorithm HOMER`
`-debug`	If included, this optional parameter will change the output produced by the program, increasing the information printed to the console	`-debug`

9.4.3 The RunMLClassifier Utility

The main drawback in using MULAN is the need to have some Java language expertise. The authors of this book provide in the companion repository an utility, named `RunMLClassifier`, aimed to help users lacking this competence. It is a Java program which eases running many of the classifiers implemented in MULAN. The source code of this program, along the specific versions of the MULAN and WEKA libraries and a compilation script, can be found in the `RunMLClassifier` folder of the repository.

To run the `RunMLClassifier` program, assuming the user is located in the same directory that the `.jar` file is obtained once it is compiled, the following sentence[5] has to be entered at the command line. The meaning of each parameter is detailed in Table 9.1.

```
java -jar RunMLClassifier.jar -path P -dataset D -folds
F -algorithm A [-debug]
```

---

[5] Although due to the page width limit the sentence appears in the text divided into two lines, it has to be entered as only one.

Table 9.2 Valid values for the `-algorithm` parameter of the RunMLClassifier utility

Value	MULAN class instantiated as classifier
BPMLL	`BPMLL()`
BR-J48	`BinaryRelevance(new J48())`
BRkNN	`BRkNN(10)`
CC-J48	`ClassifierChain(new J48())`
CLR	`CalibratedLabelRanking(new J48())`
ECC-J48	`EnsembleOfClassifierChains(new J48(), 10, true, false)`
EPS-J48	`EnsembleOfPrunedSets(80, 10, 0.2, 2,` `PrunedSets.Strategy.values()[0], 2, new J48())`
HOMER	`HOMER(new BinaryRelevance(new J48()), (numLabels < 4 ?` `numLabels : 4), Method.Random)`
IBLR-ML	`IBLR_ML()`
LP-J48	`LabelPowerset(new J48())`
MLkNN	`MLkNN(10, 1.0)`
PS-J48	`PrunedSets(new J48(), 2,` `PrunedSets.Strategy.values()[0], 2)`
RAkEL-BR	`RAkEL(new BinaryRelevance(new J48()))`
RAkEL-LP	`RAkEL(new LabelPowerset(new J48()))`

The utility instances each classifier using default or recommended values. As can be seen, only one classifier can be chosen to be run over the MLD partitions. The `RunMLClassifier` is designed to be launched independently, maybe in parallel, for each algorithm the user is interested in. An example run with this utility is shown in Fig. 9.34. Without the `-debug` parameter, only the final line which summarizes the average results and standard deviations would be printed.

> MEKA and MULAN are the two main frameworks to conduct multilabel experiments, since they provide reference implementations of many MLC methods. The former has a GUI which eases the design of such experiments, while the latter only can be used while writing some code. The `RunMLClassiier` utility aims to make the use of MULAN more comfortable, establishing the methods and datasets to process through command line parameters, instead of writing and compiling a Java program.

9.5 Summarizing Comments

Learning from multilabeled data is a quite challenging task. Datasets of this kind come in diverse file formats and are distributed among a few Web repositories. Once the data files have been obtained, they have to be imported to the learning tool, some

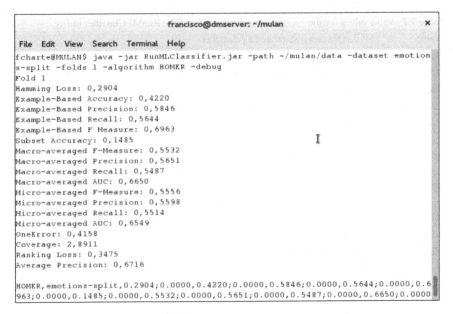

```
                              francisco@dmserver: ~/mulan                            ×

 File  Edit  View  Search  Terminal  Help
fcharte@MULAN$ java -jar RunMLClassifier.jar -path ~/mulan/data -dataset emotion
s-split -folds 1 -algorithm HOMER -debug
Fold 1
Hamming Loss: 0,2904
Example-Based Accuracy: 0,4220
Example-Based Precision: 0,5846
Example-Based Recall: 0,5644
Example-Based F Measure: 0,6963
Subset Accuracy: 0,1485                                    I
Macro-averaged F-Measure: 0,5532
Macro-averaged Precision: 0,5651
Macro-averaged Recall: 0,5487
Macro-averaged AUC: 0,6650
Micro-averaged F-Measure: 0,5556
Micro-averaged Precision: 0,5598
Micro-averaged Recall: 0,5514
Micro-averaged AUC: 0,6549
OneError: 0,4158
Coverage: 2,8911
Ranking Loss: 0,3475
Average Precision: 0,6716

HOMER,emotions-split,0.2904;0.0000,0.4220;0.0000,0.5846;0.0000,0.5644;0.0000,0.6
963;0.0000,0.1485;0.0000,0.5532;0.0000,0.5651;0.0000,0.5487;0.0000,0.6650;0.0000
```

Fig. 9.34 The utility trains and evaluates the MLD partitions with the specified classifier

times with a preliminary conversion step. A large set of multilabel characterization metrics exist in the literature, most of them were described in Chap. 3, and dozens of methods have been defined to preprocess and classify multilabel data, as is shown in Chaps. 4 to 8.

In this chapter, the Web sites from which the MLDs can be downloaded, as well as the existing file formats, have been thoroughly detailed. Tools such as the mldr.datasets R package, along with the RUMDR repository, can automate most of the tasks associated with MLDs, including getting, citing, partitioning, and exporting them to several learning software frameworks. These MLDs can be analyzed by means of the functionality found in MEKA and the mldr R package. Both provide EDA tools, comprehensively explained through this chapter.

The last section has been focused on the use of tools aimed at conducting multilabel experiments, specifically MEKA and MULAN. Although there are a few multilabel algorithms implemented outside of these two frameworks, currently they are the most prominent and widely used.

References

1. Tsoumakas, G., Xioufis, E.S., Vilcek, J., Vlahavas, I.: MULAN: A Java library for multi-label learning. J. Mach. Learn. Res. **12**, 2411–2414 (2011)
2. Read, J., Reutemann, P.: MEKA multi-label dataset repository. http://sourceforge.net/projects/meka/files/Datasets/

3. Chang, C.C., Lin, C.J.: Libsvm: a library for support vector machines. ACM Trans. Intell. Syst. Technol. **2**(3), 1–27 (2011)
4. Alcala-Fdez, J., Fernández, A., Luengo, J., Derrac, J., García, S., Sánchez, L., Herrera, F.: KEEL data-mining software tool: data set repository and integration of algorithms and experimental analysis framework. J. Mult-Valued Log. Soft Comput. **17**(2–3), 255–287 (2011)
5. Charte, F., Charte, D., Rivera, A.J., del Jesus, M.J., Herrera, F.: R Ultimate Multilabel Dataset Repository. https://github.com/fcharte/mldr.datasets
6. Tsoumakas, G., Xioufis, E.S., Vilcek, J., Vlahavas, I.: MULAN multi-label dataset repository. http://mulan.sourceforge.net/datasets.html
7. Chang, C.C., Lin, C.J.: LIBSVM data: multi-label classification repository. http://www.csie. ntu.edu.tw/~cjlin/libsvmtools/datasets/multilabel.html
8. Alcala-Fdez, J., Fernández, A., Luengo, J., Derrac, J., García, S., Sánchez, L., Herrera, F.: KEEL multi-label dataset repository. http://sci2s.ugr.es/keel/multilabel.php
9. Charte, F., Charte, D., Rivera, A.J., del Jesus, M.J., Herrera, F.: R ultimate multilabel dataset repository. In: Proceeedings of 11th International Conference on Hybrid Artificial Intelligent Systems, HAIS'16, vol. 9648, pp. 487–499. Springer (2016)
10. Bhatia, K.H., Jain, P.J., Varma, M.: The extreme classification repository: multi-label datasets & code. http://research.microsoft.com/en-us/um/people/manik/downloads/XC/XMLRepository.html
11. R Core Team: R: A language and environment for statistical computing. R Foundation for Statistical Computing, Vienna, Austria (2014). http://www.R-project.org/
12. Tomás, J.T., Spolaôr, N., Cherman, E.A., Monard, M.C.: A framework to generate synthetic multi-label datasets. Electron. Notes Theor. Comput. Sci. **302**, 155–176 (2014)
13. Read, J., Pfahringer, B., Holmes, G.: Generating synthetic multi-label data streams. In: Proceedings of European Conference on Machine Learning and Principles and Practice of Knowledge Discovery in Databases, ECML PKDD'09, pp. 69–84 (2009)
14. Charte, F., Charte, D.: Working with multilabel datasets in R: the mldr package. R J. **7**(2), 149–162 (2015)

Glossary

5-fcv/10-fcv   5/10-fold cross-validation.

Attribute   Each one of the columns in an instance.

Bag   Collection of physical data instances making up a logical sample in multi-instance classification.

Bipartition   Structure of the output produced by many MLC classifiers, stating which labels are relevant or not to the processed instance.

Cardinality   Average number of active labels per instance in a multilabel dataset.

Clustering   Technique aimed to discover how similar data points can be assembled into groups.

Dataset   A collection of instances. Can be seen as a matrix with a set of rows (instances), each one with a set of columns (attributes).

Density   Label density is a dimensionless metric derived from *cardinality*.

Dimensionality   Usually referred to the number of input variables a dataset has.

Diversity   Label diversity is a metric to measure the different label combinations or labelsets that a set of labels have generated in a multilabel dataset.

Ensemble   Set of learners along with a strategy to join their predictions.

Feature   See *attribute*.

Feature engineering   Diverse techniques to select and extract features from a dataset to reduce dimensionality.

Feature selection   Technique to choose the most relevant features from a dataset.

Imbalance   Prominent inequality in the frequency each class appears in a dataset.

Imbalance ratio   Proportion between the majority and minority classes in a dataset.

Input space   The space generated by the attributes used as predictors in a dataset.

Instance   A data point (row) defined by a set of values (columns).

Labelset   Vector of output values associated with each instance in a multilabel dataset.

Lazy method   A DM method that does not generate a model and that defers the work until a new instance arrives.

Majority class/label   The most frequent class or a frequent label in a dataset.

Minority class/label   The least frequent class or a rare label in a dataset.

© Springer International Publishing Switzerland 2016
F. Herrera et al., *Multilabel Classification*,
DOI 10.1007/978-3-319-41111-8

Neuron   Each one of the units which conform a layer in a neural network.

Outlier   Data instance whose attributes have some values out of the common range but cannot be considered as noise.

Output space   The space generated by the labels (output attributes) in a multilabel dataset.

Oversampling   Resampling technique that produces additional data samples.

Preprocessing   Family of tasks aimed to perform data cleaning and selection of relevant data.

Repository   Resource where datasets and sometimes software are provided. It is usually a Web site.

Resampling   Technique used to take subsets of the original samples and remove samples to add samples.

Sample   See *instance*.

Segmentation   The usual process to extract features from signal information such as images and sounds.

Supervised   Supervised methods are guided by the labels associated with data samples.

Tree   A knowledge representation model from which usually simple rules can be extracted.

Undersampling   Resampling techniques that remove data samples from a dataset.

Unsupervised   Unsupervised methods only work with input features, without being guided by class labels.

Printed in the United States
By Bookmasters